128 Advances in Polymer Science

Springer
Berlin
Heidelberg
New York
Barcelona
Budapest
Hong Kong
London
Milan
Paris
Santa Clara
Singapore
Tokyo

Polymer Analysis
Polymer Physics

With contributions by
A. L. Andrady, B. Narasimhan, J.-P. Pascault, H. Pasch,
N. A. Peppas, B. A. Rozenberg, R. J.J. Williams

With 116 Figures and 9 Tables

Springer

This series presents critical reviews of the present position and future trends in modern polymer research. It is addressed to all polymer and material scientists in industry and the academic community who wish to keep abreast of advances in the topics covered.

As a rule, contributions are specially commissioned. The editors and publishers will, however, always be pleased to receive suggestions and supplementary information. Papers are accepted for "Advances in Polymer Science" in English.

In references "Advances in Polymer Science" is abbreviated Adv. Polym. Sci. and is cited as a journal.

Springer WWW homepage:http://www.springer.de

ISBN 3-540-61218-1 Springer-Verlag Berlin Heidelberg New York
ISBN 0-387-61218-1 Springer-Verlag New York Berlin Heidelberg

© Springer-Verlag Berlin Heidelberg New York 1997
ISSN 0065-3195
Printed in Germany

Typesetting: Macmillan India Ltd., Bangalore-25
SPIN: 10508864   02/3020 - 5 4 3 2 1 0 - Printed on acid-free paper

Editors

Table of Contents

Analysis of Complex Polymers by Interaction Chromatography

Harald Pasch

Deutsches Kunststoff-Institut, Schloßgartenstr. 6, 64289 Darmstadt, Germany

This review presents the principle ideas of using interaction chromatography for analyzing different heterogeneities in complex polymers. Based on the discussion of a unified mechanism of chromatography of polymers, it is demonstrated that, by selecting the corresponding chromatographic conditions, in a number of cases it is possible to determine heterogeneity parameters separately from each other. Emphasizing the usefulness of liquid chromatography at the critical point of adsorption, it is shown that functional homopolymers can be analyzed solely with respect to functionality. Binary polymer blends are separated into their components regardless of the molar mass distribution. Depending on their particular structure, copolymers can be analyzed efficiently by gradient HPLC or critical mode chromatography. Finally, further developments including two-dimensional techniques are addressed.

List of Symbols and Abbreviations

CC	liquid chromatography at the critical point of adsorption
CH	chemical heterogeneity
CoStMMA	poly(styrene-*co*-methyl methacrylate)
ε	interaction energy of the monomer segment with the stationary phase
ε_c	critical energy of adsorption
ELSD	evaporative light scattering detector
EO	ethylene oxide
EPDM	ethylene propylene diene terpolymer
FTD	functionality type distribution
FTIR	Fourier transform infrared spectroscopy
ΔG	Gibbs free energy
ΔH	change in enthalpy
HPLC	high performance liquid chromatography
K_{Ads}	distribution coefficient of real LAC
K_d	distribution coefficient
K_{el}	distribution coefficient of electrostatic interactions
K_{Excl}	distribution coefficient of real SEC
K_{LAC}	distribution coefficient of ideal LAC
K_{SEC}	distribution coefficient of ideal SEC
LAC	liquid adsorption chromatography
MALDI-MS	matrix-assisted laser desorption/ionization mass spectrometry
MAS	methacrylic acid
MEK	methylethylketone
MMD	molar mass distribution
n_A	number of moles of species A
n_B	number of moles of species B
P	degree of polymerization
PBA	poly(butyl acrylate)
PBMA	poly(butyl methacrylate)
PCHMA	poly(cyclohexyl methacrylate)
PDMA	poly(decyl methacrylate)
PEA	poly(ethyl acrylate)
PEG	poly(ethylene glycol)
PEHA	poly(ethylhexyl acrylate)
PEMA	poly(ethyl methacrylate)
PEO	poly(ethylene oxide)
PMA	poly(methyl acrylate)
PMMA	poly(methyl methacrylate)
PO	propylene oxide
PPE	polyphenylene ether
PPO	poly(propylene oxide)

PS	polystyrene
PVC	poly(vinyl chloride)
ΔS	change in entropy
SAN	poly(styrene-*co*-acrylonitrile)
SEC	size exclusion chromatography
SFC	supercritical fluid chromatography
St	styrene
T	temperature
THF	tetrahydrofurane
TO	trioxocane
V_o	interparticle volume
V_p	pore volume
V_R	retention volume
V_{stat}	volume of stationary phase
VA	vinyl acetate

1 Introduction

Synthetic polymers are produced by chain polymerization or step growth polymerization. Due to differences in the lifetime of activated species or the size and reactivity of the oligomers which are coupled in each reaction step, synthetic polymers are heterogeneous in molar mass. Copolymers are produced from more than one monomer species. In general, the different monomer species are differently incorporated in the polymer chain which causes distribution in chemical composition. Distributions in molar mass and chemical composition are also to be expected in polymers derived from homopolymers by incomplete chemical modifications, e.g. in partially hydrolyzed poly(vinyl acetate) [1].

In the synthesis of tailor-made polymers, i.e. polymers with required properties, telechelics or macromonomers are frequently used. These oligomers or polymers usually contain functional groups at the polymer chain end. Depending on preparation procedure, they can have a different number of functional endgroups, i.e. be mono-, bifunctional etc. In addition, polymers can have different architectures, i.e. they can be branched (star- or comb-like), and they can be cyclic.

The different heterogeneities, summarized in the term "molecular heterogeneity", can be superimposed one on another, i.e. bifunctional molecules can be linear or branched, linear molecules can be mono- or bifunctional, copolymers can be block or graft copolymers etc. – see Fig. 1. In order to characterize heterogeneous polymers, i.e. polymers which are distributed in more than one property, it is necessary to know a set of functions $f_i(M)$, e.g. the molar mass distributions within each other type of heterogeneity. Clearly, it is very difficult in a general case to solve this characterization problem.

Using the traditional methods of polymer analysis, such as infrared spectroscopy or nuclear magnetic resonance, one is able to determine the type of monomers or functional groups, present in the sample. However, the determination of functional endgroups is complicated for long molecules because of low concentration. On the other hand this type of determination does not yield information on how functional groups or different monomer species are distributed in the polymer molecule. With respect to methods sensitive to the size of the macromolecule, one can encounter other difficulties. Size exclusion chromatography, which is most frequently used to separate polymer molecules from each other according to their molecular size in solution, must be used very carefully when analyzing heterogeneous polymers (telechelics, block and graft copolymers). The dimensional distribution of macromolecules can in general be unambiguously correlated with molar mass distribution only within one heterogeneity type. For samples consisting of a mixture of molecules of different functionality, the distribution obtained represents an average of dimensional distributions of molecules having a different functionality and, therefore, cannot be attributed to a specific functionality type without additional assumptions [2].

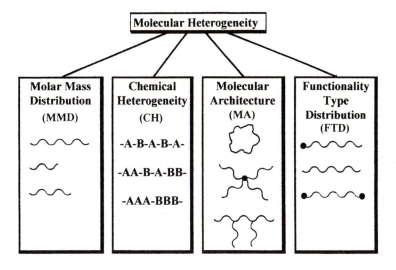

Superposition of Heterogeneities

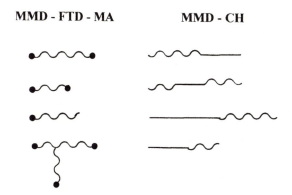

Fig. 1. Schematic representation of the molecular heterogeneity of polymers

For many years detection in SEC has been emphasized over fractionation. Reliance has been placed upon detection to reveal copolymer composition distribution information in addition to the molar mass distribution. The inadequacy of this approach for analysis of complex polymers can be explained with reference to Fig. 2 [3]. For a linear homopolymer distributed only in molar mass, fractionation by SEC results in one molar mass being present at each retention volume. The polymer at each retention volume is monodisperse. If a polymer blend of two linear homopolymers is fractionated, then at most two different molar masses can be present at one retention volume. However, when a linear copolymer is fractionated, a multitude of different combinations of molar mass, composition and sequence length can combine to give the same

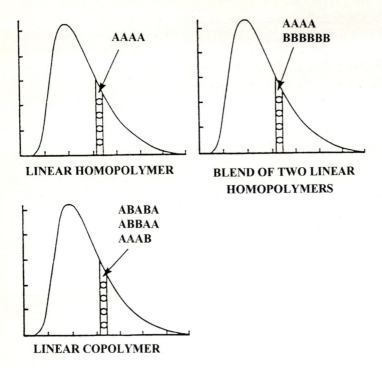

Fig. 2. SEC fractionation showing detector contents at a given elution volume. (From [3] with permission)

molecular size in solution. In this case fractionation with respect to molecular size is completely ineffective in assisting the analysis of composition as well as molar mass distribution.

Dimensional heterogeneity is thus seen to be closely related to all other types of molecular heterogeneity. Not knowing the dimensional distribution, one cannot analyze the chemical composition or functionality type distribution, and not knowing these heterogeneities, one cannot unambiguously interpret SEC data.

For the determination of the dimensional heterogeneity of complex polymers, several methods have been proposed in recent years. In particular, multiple detection SEC systems using n independent detector signals allow the composition calculation of n components in the sample. SEC with viscometric detection permits the on-line measurement of Mark-Houwink coefficients for copolymers of various architectures, and SEC with light scattering detection yields copolymer molar masses directly and independently of their structure. Another approach is the use of a chromatographic method which is sensitive to size, chemical composition and functionality. Liquid adsorption chromatography, or more generally speaking "interaction" chromatography (including adsorption, hydrophobic interactions, critical point phenomena), where the macromolecules interact with the stationary phase, is sensitive to both size and local structure of the macromolecule. Interaction chromatography is being used to fractionate copolymers with respect to composition [4–6]. However, in addition to the effect of composition, interference from size exclusion effects and other property distributions must be taken into account.

This review presents the principal ideas of using interaction chromatography for analyzing different heterogeneities in complex polymers. Based on the discussion of a unified mechanism of chromatography of polymers it will be demonstrated that, by selecting the corresponding chromatographic conditions, in a number of cases it is possible to investigate heterogeneities separately from each other. It will be illustrated that functional homopolymers may be analyzed exclusively with respect to functionality. Block copolymers may be separated with respect to the lengths of the individual blocks, and binary polymer blends may be separated into their components regardless of the molar mass distribution. Finally, two-dimensional chromatographic techniques will be reviewed.

2 Theoretical Background

Any chromatographic process relates to the selective distribution of an analyte between the mobile and the stationary phase of a given chromatographic system. In liquid chromatography the solvent, with a volume V_o in the interparticle space, moving along the column at a certain velocity, is the mobile phase and the porous adsorbent, having a pore volume V_p, is the stationary phase. The distribution coefficient K_d equals the ratio of the concentrations of the analyte in the stationary and the mobile phases. In classical SEC a distribution of the analyte between the interparticle volume and the accessible pore volume takes place and the retention volume V_R is determined by

$$V_R = V_o + V_p K_d. \tag{1}$$

K_d is related to the change in Gibbs free energy ΔG at the point where the analyte molecules pass from the mobile into the stationary phase:

$$\Delta G = \Delta H - T\Delta S = - RT \ln K_d \tag{2}$$

$$K_d = \exp(\Delta S/R - \Delta H/RT). \tag{3}$$

The change in Gibbs free energy may be due to different effects:

1. inside the pore, which is limited in dimensions, the macromolecule can not occupy all possible conformations and, therefore, the conformational entropy ΔS decreases;

2. when penetrating the pores, the macromolecule may interact with the pore walls resulting in a change in enthalpy ΔH.

Therefore, depending on the magnitude of entropic and enthalpic effects, different cases in liquid chromatography of polymers must be considered [1]. In *ideal SEC*, separation is exclusively directed by conformational changes of the macromolecules and ΔH by definition is zero:

$$K_{SEC} = \exp(\Delta S/R). \tag{4}$$

As the conformational entropy decreases ($\Delta S < 0$) the distribution coefficient of ideal SEC is $K_{SEC} < 1$. The maximum value, $K_{SEC} = 1$, is related to zero change in conformational entropy, i.e. to a situation where all pores are accessible to the analyte molecules (separation threshold). At $K_{SEC} = 0$, the analyte molecules are too large to penetrate the pores (exclusion limit). Accordingly, the separation range is $0 < K_{SEC} < 1$. The retention volume for ideal SEC equals

$$V_R = V_o + V_p K_{SEC}. \tag{5}$$

If enthalpic effects due to electrostatic interactions of the macromolecules and the pore walls have to be taken into account, the distribution coefficient K_{Excl} of *real SEC* is

$$K_{Excl} = \exp(\Delta S/R - \Delta H/RT)$$
$$= \exp(\Delta S/R)\exp(-\Delta H/RT) = K_{SEC}K_{el}. \tag{6}$$

In this case the retention volume is a function of K_{SEC} and K_{el}. If electrostatic interactions occur at the outer surface of the stationary phase as well, an additional term $V_{stat}K_{LAC}$ has to be accounted for.

As for adsorption chromatography (LAC), where separation is directed by adsorptive interactions between the macromolecules and the stationary phase, an ideal case may be defined as well. In *ideal LAC* conformational changes are assumed to be zero ($\Delta S = 0$) and the distribution coefficient is exclusively determined by enthalpic effects:

$$K_{LAC} = \exp(-\Delta H/RT). \tag{7}$$

Depending on the pore size of the stationary phase, two possible cases have to be discussed.

1. For small pore stationary phases, separation occurs exclusively on the outer surface. The pores are not accessible to the macromolecules ($K_{SEC} = 0$). Accordingly, the retention volume is a function of the interparticle volume and the volume of the stationary phase (V_{Stat}):

$$V_R = V_o + V_{Stat}K_{LAC}. \tag{8}$$

2. Without accessibility restrictions of the pores of the stationary phase ($K_{SEC} = 1$), the pore volume V_p adds to the interparticle volume:

$$V_R = V_o + V_p + V_{Stat}K_{LAC}. \tag{9}$$

In *real LAC* only a fraction of the pores of the stationary phase is accessible and, therefore, entropic interactions must be assumed. Accordingly, the distribution coefficient is a function of ΔH and ΔS (compare to real SEC):

$$K_{Ads} = \exp(\Delta S/R - \Delta H/RT). \qquad (10)$$

The retention volume now is a function of enthalpic interactions at the surface of the stationary phase, entropic effects due to the limited dimensions of the pores, and possible enthalpic interactions inside the pores (K_{el}). Therefore the expression for V_R in real LAC is formally similar to that in real SEC:

$$V_R = V_o + V_p(K_{SEC}K_{el}) + V_{Stat}K_{LAC}. \qquad (11)$$

In conclusion, ideal SEC and ideal LAC are extremes of the same principle. In the real SEC and LAC modes, retention responds to both enthalpic and entropic interactions, and only the predominance of one of these interactions decides which mode is operating. Accordingly, real SEC and real LAC can be regarded as mixed mode chromatographic methods. With chemically heterogeneous polymers, effects are even more dramatic because exclusion and adsorption act differently on molecules of different composition.

In a more general sense, the size exclusion mode of liquid chromatography relates to a separation regime, where entropic interactions are predominant and $T\Delta S > \Delta H$. In the reverse case, $\Delta H > T\Delta S$, separation is mainly directed by enthalpic interactions. As both separation modes in the general case are affected by the macromolecule size and the pore size, a certain energy of interaction ε may be introduced, characterizing the specific interactions of the monomer unit of the macromolecule and the stationary phase. ε is a function of the chemical composition of the monomer unit, the composition of the mobile phase of the chromatographic system, and the temperature.

The theory of adsorption at porous adsorbents predicts the existence of a finite critical energy of adsorption ε_c, where the macromolecule starts to adsorb at the stationary phase. Thus, at $\varepsilon > \varepsilon_c$ the macromolecule is adsorbed, whereas at $\varepsilon < \varepsilon_c$ the macromolecule remains unadsorbed. At $\varepsilon = \varepsilon_c$ the transition from the unadsorbed to the adsorbed state takes place, corresponding to a transition from one to another separation mechanism. This transition is termed "critical point of adsorption" and relates to a situation, where the adsorption forces are exactly compensated by the entropy losses $T\Delta S = \Delta H$ [2, 7]. Accordingly, at the critical point of adsorption the Gibbs free energy is constant ($\Delta G = 0$) and the distribution coefficient is $K_d = 1$, irrespective of the molar mass of the macromolecules. The critical point of adsorption relates to a very narrow range between the size exclusion and adsorption modes of liquid chromatography. It is, therefore, very sensitive towards temperature and mobile phase composition.

The transition from one to another chromatographic separation mode by changing the temperature or the composition of the mobile phase for the first time was reported by Tennikov et al. [8] and Belenkii et al. [9, 10]. They showed that a sudden change in elution behaviour may occur by small variations in the

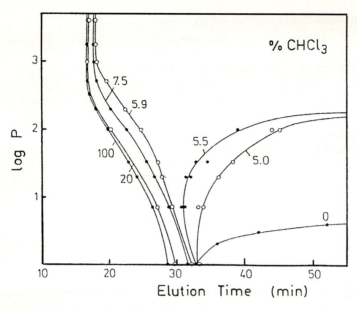

Fig. 3. Elution of polystyrene as a function of the degree of polymerization P and the eluent composition: stationary phase silica gel; mobile phase chloroform-carbon tetrachloride. (From [8] with permission)

solvent strength (see Fig. 3 for the behavior of polystyrene in a mobile phase of chloroform-carbon tetrachloride). In eluents comprising more than 5.9% chloroform, conventional SEC behaviour is obtained, that is elution time increases with decreasing molar mass of the sample. At lower concentrations of chloroform (5.5% and less) a completely reversed retention behavior is obtained. The increase of the elution time with increasing molar mass indicates that the adsorption mode is now operating. Thus, just by gradually changing the eluent composition, a transition from the SEC to the LAC mode and vice versa may be achieved. The point of transition from SEC to LAC is the critical point of adsorption, and chromatographic separations at this point are termed "liquid chromatography at the critical point of adsorption".

The use of liquid chromatography at the critical point of adsorption (CC, critical chromatography) for the determination of the functionality type distribution of telechelics was demonstrated by Evreinov and co-workers [2, 11–15]. Meanwhile, a significant number of investigations on functional polyolefines, oligoethers, polyesters, and epoxy resins were conducted showing the usefulness of this new technique.

A theoretical approach to the determination of block copolymers of the $A_n B_m$ type has been developed by Gorbunov and Skvortsov [17, 18]. They proposed the term "chromatographic invisibility" to describe the insensitivity of K_d to the molar mass at the critical point of adsorption. According to this approach a block copolymer may be separated with respect to the length of the "visible" block by conducting the experiment under conditions where one block is chromatographically "invisible". These are critical conditions of this particular block. In general, the types of macromolecules which may be investigated by critical chromatography regardless of the pore size of the stationary phase are summarized in [18]. Experimental evidence for the validity of this approach, however, is still limited to a few investigations [19, 20]. Recent investigations on the quantitative determination of the molar mass distribution of single blocks in block copolymers will be dealt with in the Sect. 4.2.

3 Instrumentation

3.1 General Considerations

The typical instrument used for interaction chromatography is very similar to a conventional SEC instrument. The sample is dissolved in a solvent, injected into a flow stream of the eluent at the top of the chromatographic column, and carried through the column at a constant flow rate. The columns are filled with a non-porous or porous packing material, which may interact with the solute molecules due to a certain surface activity. Upon leaving the column, the solute molecules enter one or more sequentially attached detectors. In SEC a typical detector is a differential refractometer, but many different detectors may be used.

Interaction chromatography utilizes columns that are intended to encourage adsorption and partition mechanisms. As will be shown later, at one column, depending on the composition of the mobile phase, simultaneous adsorption, partition, and size exclusion may occur. In interaction chromatography, column packings with a hydrophobic surface are very frequently used. They are termed "reversed-phase" columns because the packings are less polar than the mobile phase.

Because of the different separation mechanisms employed in interaction chromatography, the solvents used are usually different from the solvents in SEC. Binary or ternary mixtures of organic and aqueous solvents are frequently used as the mobile phase. The ratio of the solvents may be constant during the chromatographic run (isocratic elution) or may be programmed to vary with time (gradient elution).

3.2 Stationary and Mobile Phases

In interaction chromatography the packing material or stationary phase has active sites at the surface, where interaction with the solute molecules takes place according to their polarity. In general, the stationary phase must withstand high pressure and, in gradient elution, the particle and pore size must not change when the eluent composition varies. Therefore, stationary phases of bare or modified silica are usually preferred, although crosslinked poly(acrylonitrile) or polystyrene may be used as well [1]. Criteria for the quality of stationary phases are size and shape of the particles, specific surface, pore size and pore size distribution, chemical stability, and pressure stability. In general, stationary phases may be classified with respect to their surface polarity (Table 1).

As already mentioned, silica gel is the most important stationary phase in interaction chromatography. Its surface behaves like a weak acid, and retention is directed by adsorption effects. The chemical modification of silica gel alters the surface by introduction of functional groups. A transformation of the polar surface to a less polar surface usually takes place, and retention now is directed by hydrophobic interactions. However, polar functional groups may be attached to the silica surface via a hydrophobic spacer (cyano, diol, amino phases). These stationary phases exhibit hydrophilic *and* hydrophobic properties and may be used in normal-phase as well as in reversed-phase chromatography.

For the selection of mobile phases the following criteria must be observed: solubility of the sample, controlled interactions with the solute and the stationary phase, suitability for a specific detector, chemical stability, miscibility as a precondition for gradient techniques, solubility of buffers and modifiers, viscosity, and environmental compatibility. The classification of mobile phases in interaction chromatography is usually done with respect to the solvent strength or the elution power, and the so called "eluotropic series" compares different solvents in the order of increasing solvent strength ε°. The eluotropic series must be determined experimentally for each stationary phase by determining the retention time as a function of the composition of the eluent. For stationary phases of different polarity the eluotropic series will be completely different. Thus, for silica gel as the stationary phase n-heptane exhibits the lowest elution power ($\varepsilon^\circ = 0$), whereas i-propanol is a strong eluent ($\varepsilon^\circ = 0.82$). In reversed-phase chromatography water is the weakest eluent, and n-heptane exhibits a high elution power.

Table 1. Classification of stationary phases in interaction chromatography

non-polar	medium-polar	polar
styrene-divinylbenzene	cross-linked hydroxyethyl methacrylate	agarose silica gel aluminium oxide
bonded phases		
X-n-octadecyl	X-cyanopropyl	X-nitrophenyl
X-n-octyl	X-diol	X-aminopropyl
X-phenyl		
(X-silica gel)		

3.3 Detector Requirements

Once the chromatographic separation on the column has been conducted, the composition of the eluent at the column end must be determined using a detector. In all HPLC detectors, the eluent flows through a measuring cell where the change of a physical or chemical property with elution time is detected. The most important parameter of the detector is sensitivity, which is influenced by the noise and baseline drift, the absolute detection limit of the detector, the linearity, the detector volume (band broadening), and the effects of pressure, temperature and flow (pulsation, gas bubbles).

In gradient elution, detection requires quantitative measurement of the sample components in an eluent whose composition and, hence, physical properties alter in the course of the analysis. The detection problem in gradient elution can be solved by either using a selective detector sensitive to a property of only the solute or stripping off the solvent with subsequent measurement of non-volatile residues.

The current status and further prospects of HPLC detectors have been dealt with in detail in a number of reviews [21–23]. In brief, universal and selective detectors are used in liquid chromatography. With universal detectors the solute concentration in the eluate may be detected regardless of the chemical structure of the solute. The most commonly used universal detector is the differential refractometer, which measures changes of the refractive index of the polymer solution as compared to the refractive index of the pure solvent. This detector can be used only in isocratic elution. The change in composition of the eluent in gradient elution and the corresponding change in the refractive index of the eluent causes a significant baseline drift. In gradient elution the evaporative light-scattering detector (ELSD) can be used as a universal detector despite response linearity problems. Any non-volatile solute in every eluent which can be vaporized under operating conditions may be detected. The eluate from the HPLC column is directed into a nebulizer where a gas supply atomizes the liquid stream. The droplets are sprayed into the evaporator where the solvent is vaporized. The resulting aerosol is then measured in a light scattering device.

The most commonly used selective detectors are photometers for UV and visible light. The accessible wavelength range is 190–800 nm, but, depending on the light absorption properties of the solute and the eluent, specific wavelengths are usually selected. A further new development, the photo diode array detector, is capable of monitoring the complete UV spectrum during the chromatographic run. This way, additional information on the structure of the sample components may be obtained. Even more detailed structural information may be obtained by using an infrared detector. However, absorption bands of the eluent may disturb the detection. In a new approach, a deposition technique is applied, removing the eluent from the sample and enabling the complete IR spectra of the sample components to be measured [24, 25]. Of particular interest for polymer analysis may be molar mass sensitive detectors (viscosity, light scattering detectors) [26–28].

4 Separation of Complex Polymers by Interaction Chromatography

The following section will describe the principal treatment of different types of molecular heterogeneity. Emphasis will be on the analysis of functional homopolymers (telechelics, macromonomers), copolymers (statistical, block, graft), and polymer blends. A detailed description of all experimental steps will be given for one or two representative examples of each group. Information on the application of the described procedures to different heteropolymer structures will be summarized briefly in the Sect. 5.1.

4.1 Analysis of Functional Homopolymers

Oligomers and polymers with reactive functional groups have been used extensively to prepare a great variety of polymeric materials. In many cases the behavior of these functional homopolymers is largely dependent on the nature and number of functional groups. In a number of important applications the functional groups are located at the end of the polymer chain; macromolecules with terminal functional groups are usually termed "telechelics" or "macromonomers". To characterize them it is necessary to have information not only about their molar mass but also on their functionality.

When functional homopolymers are synthesized, in addition to macromolecules of required functionality, functionally defective molecules are formed (see Fig. 4). For example, if a target functionality of $f = 2$ is required, then in the normal case species with $f = 1$, $f = 0$ or higher functionalities are formed as well [7]. Deviation of the average functionality from the pre-assigned one may result in a decreased or increased reactivity, cross-linking density, surface activity etc. Each functionality fraction has its own molar mass distribution. Therefore, to fully describe the chemical structure of a functional homopolymer, the determination of the molar mass distribution (MMD) and the functionality type distribution (FTD) is required.

The use of liquid chromatography for the determination of FTD of oligomers was proposed by Muenker and Hudson [29] and Evreinov [30]. Using mixtures of solvents with increasing solvent strength, hydrogenated oligobutadienes with hydroxyl and carboxyl terminal groups were separated according to their functionality types. A practically complete separation of non-, mono- and bifunctional fractions could be achieved [29]. For poly(diethylene glycol adipate) it was shown that on silica gel a separation according to the number of terminal hydroxyl groups takes place [30].

With respect to the chromatographic system, three different cases can be encountered (see Fig. 5) as follows.
1. The functional groups do not manifest themselves chromatographically. The calibration curves for fractions of different functionality coincide, and the molar

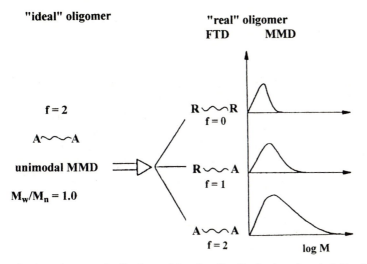

Fig. 4. Molar mass distribution and functionality distribution of a telechelic oligomer. (From [2] with permission)

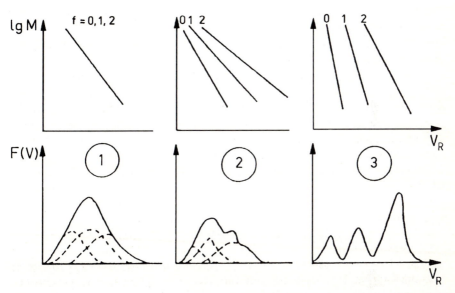

Fig. 5. Typical experimental situations in chromatography of functional homopolymers. For explanation of 1, 2, and 3 see text. (From [2] with permission)

mass distribution obtained is a superposition of the distributions of the functionality fractions (ideal SEC).

2. The functional groups manifest themselves only slightly. The calibration curves do not coincide, but because of the molar mass distributions the elution zones overlap.

3. Complete separation of the functionality fractions can be obtained, when the interaction of the terminal groups with the adsorbent is much stronger than that of the polymer chain (ideal chromatography at the critical point).

It is clear from Fig. 5 that case 3 is the preferred case because it allows the separate determination of FTD.

As was stated before, three modes of liquid chromatography may be used-size exclusion, adsorption, and critical. Depending on the chromatographic system, i.e. the pore size and the polarity of the stationary phase and the composition of the mobile phase, one of these modes will direct retention of the macromolecules. From Fig. 3 it was clear that, for a given stationary phase, even slight variations in the composition of the mobile phase may result in a sudden transition from one to another separation mode. The transition through different separation modes by changing the composition of the mobile phase is illustrated for poly(butylene glycol terephthalate) in Fig. 6 [31]. At a mobile phase of THF-heptane 80:20 v/v separation is accomplished according to the size exclusion mode. Within the same functionality first the large molecules are eluted. At THF-heptane 40:60 v/v, retention volume increases with increasing molar mass and, therefore, the adsorption mode is operating. In both separation modes an overlapping of molar mass and functionality effects is obtained. The most interesting feature is the separation at the critical point of adsorption. In this case, using a mobile phase of THF-heptane 65:35 v/v, separation of the sample into three fractions is obtained, which may be assigned to the three functionality fractions

Peaks 3–7

$$CH_3O\left[OC-\bigcirc-CO-O(CH_2)_4O\right]_n OC-\bigcirc-COOCH_3$$

Peaks 8–12

$$CH_3O\left[OC-\bigcirc-CO-O(CH_2)_4O\right]_n H$$

Peaks 13–17

$$HO(CH_2)_4O\left[OC-\bigcirc-CO-O(CH_2)_4O\right]_n H$$

At the critical point of adsorption a complete independence of V_R of the size of the macromolecules is obtained, and the separation is accomplished exclusively with respect to functionality.

One of the most important classes of functional homopolymers is the class of alkoxy and aryloxy polyethylene oxides (PEO). These oligomers and polymers are in widespread use as surfactants. Depending on their molar mass and the chemical structure of the terminal groups the amphiphilic properties change, thus influencing the surface activity. Due to the different initiation, chain transfer and chain termination mechanisms and possible impurities in the reaction mixture, species having different terminal groups bound to the PEO

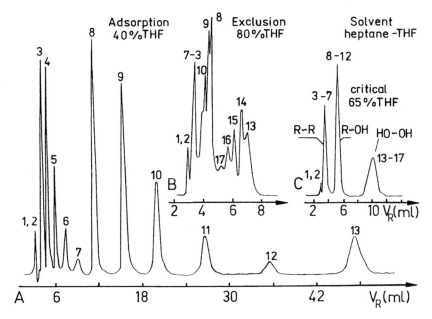

Fig. 6. Separation of hydroxy-terminated poly(butylene glycol terephthalate) as a function of the mobile phase composition: stationary phase silica gel; mobile phase: THF-heptane: peaks 1,2 solvent, peaks 3–7 nonfunctional homologues (n = 0–4), peaks 8–12 monofunctional homologues (n = 0–4), peaks 13–17 bifunctional homologues (n = 0–4). (From [2] with permission)

chain are formed. To elucidate the structure property relationships of the products, it is important to know the chemical structure and the number of the terminal groups in addition to the molar mass distribution.

It was shown by Gorshkov et al. [16] that polyethylene glycol, vinyl- and butyl-terminated PEO may be analyzed with respect to FTD by liquid chromatography at the critical point of adsorption, using a reversed-phase column RP-18 of 250 mm length, and acetonitrile-water as the eluent. For PEOs with long hydrophobic chain ends, i.e. $C_{10}H_{21}$-, $C_{12}H_{25}$- and longer, this column is not suitable, due to very strong hydrophobic interactions. To avoid irreversible adsorption on the stationary phase, packing material of lower hydrophobicity, such as RP-8 and RP-4, or very short columns must be used [32].

The critical diagram M vs retention time for a short RP-18 column of 60 mm length is shown in Fig. 7A. At acetonitrile concentrations > 47 vol.% in the solvent mixture, the retention time decreases as the molar mass of the PEO calibration sample increases; retention corresponds to a size-exclusion mode. The reverse behaviour, i.e. the adsorption mode, is obtained at acetonitrile concentrations < 45 vol.%. The critical point of adsorption is operating at an eluent composition of acetonitrile-water 46 : 54 v/v, where separation is accomplished exclusively with respect to the functional endgroups- (see Fig. 7B). For the alkoxy-terminated PEOs, two distinctively different fractions are obtained, due to the formation of polyethylene glycol as an unwanted by-product. In the case

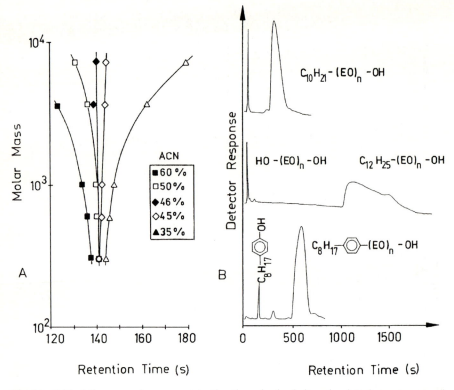

Fig. 7A. Critical diagram molar mass vs retention time of polyethylene glycol. **B** chromatograms of functional PEOs at the critical point of PEG. Stationary phase Nucleosil RP-18, 60×4 mm I.D.: eluent acetonitrile (ACN)-water. (From [32] with permission)

of the octylphenol-terminated PEO, the UV detector reveals two additional peaks which are identified as starting octylphenol and C_4H_9-terminated PEO, which might have been formed due to impurities in the starting octylphenol. In agreement with the expected behaviour for a reversed-phase column, the retention time of the samples increases with increasing hydrophobicity of the terminal group. Therefore, the elution order with respect to the terminal group is OH (PEG) $\ll C_{10}H_{21} < C_{12}H_{25}$. For $C_{13}H_{27}$- and longer terminal groups, a dramatic increase of the retention times near the critical point is obtained, suggesting strong hydrophobic interactions.

In order to decrease the hydrophobicity of the stationary phase and the retention times accordingly, an RP-8 instead of an RP-18 column was used, now allowing the separation of $C_{13}H_{27}$- and $C_{15}H_{31}$-terminated PEOs. In addition to the previously detected peaks, for the aryloxy samples a third functionality fraction was eluted, which corresponds to the α,ω-diaryloxy species. The concentration of this third functionality fraction, however, is very low and can be detected in the product only with a 30-fold increased sensitivity of the UV detector (see Fig. 8).

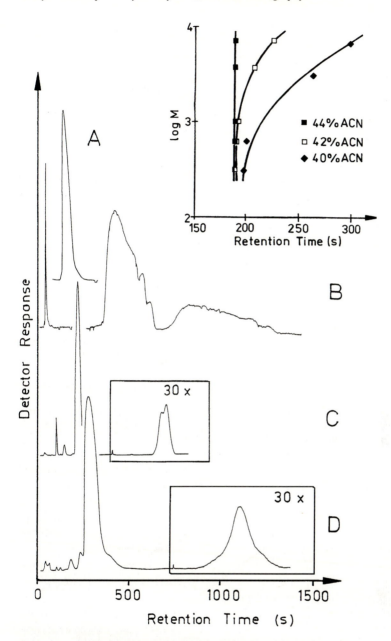

Fig. 8A–D. Critical diagram molar mass vs retention time of PEG (*insert upper right corner*) and chromatograms of functional PEOs at the critical point of adsorption of PEG: stationary phase Nucleosil RP-8, 60×4mm I.D.: eluent acetonitrile (ACN)-water 44:56 v/v: samples: **A** $C_{10}H_{21}$-PEO; **B** $C_{13}H_{27}$, $C_{15}H_{31}$-PEO; **C** octylphenol-PEO; **D**, nonylphenol-PEO. (From [32] with permission)

An additional option for the analysis of alkoxy/aryloxy PEOs with respect to functionality is the use of mixed modes of separation. Working in the SEC mode near the critical point of adsorption it is possible to separate these products both with respect to functionality and molar mass (Fig. 9). Using an

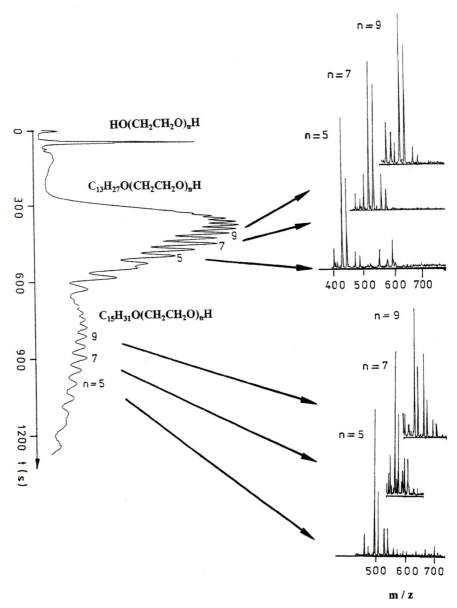

Fig. 9. Separation of a technical $C_{13}H_{27},C_{15}H_{31}$-PEO with respect to functionality and molar mass, and identification of the peaks by MALDI-MS

RP-18 column and an eluent composition of acetonitrile-water 70:30 v/v, a technical alkoxy PEO is separated into the PEG, $C_{13}H_{27}$- and $C_{15}H_{31}$-PEO fractions. In addition to the functionality separation, within the functionality fractions a separation into oligomers is obtained. After identification of the peaks by matrix-assisted laser desorption/ionization mass spectrometry (MALDI-MS), the molar masses of the individual fractions are calculated via a conventional SEC calibration [33].

4.2 Analysis of Copolymers

Copolymers are complex macromolecular systems which are formed when two or more monomers of different chemical structures react in a polymerization process. As a result products are obtained which, in addition to the MMD, are characterized by a certain chemical composition. This chemical composition may be expressed by an average number, characterizing the total amount of each monomer in the reaction product. More detailed information, however, is obtained when the chemical heterogeneity (CH) is determined, characterizing the sequence distribution of the different monomer units along the polymer chain.

In general, when reacting two monomers in a polymerization reaction, different products may be formed, depending on the reactivity ratios of the monomers and the reaction conditions (see Fig. 10). Therefore, in order to describe the reaction products fully it is necessary to analyze the copolymer (MMD, CH) and to determine the amount of homopolymers which may have been formed as unwanted by-products. For a block copolymer it would be of

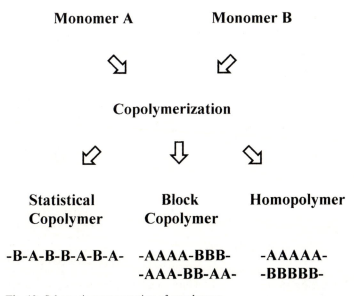

Fig. 10. Schematic representation of copolymers

particular interest not only to determine the total MMD but also the MMD of the single blocks.

The analysis of various copolymers by gradient HPLC and chromatographic cross-fractionation has been extensively discussed in the excellent monography of Glöckner [1]. In brief, the complex molar mass-chemical composition distribution of copolymers requires separation in more than one direction. The classical approach is based upon the dependence of copolymer solubility on composition and chain length. A solvent/non-solvent combination fractionating solely by molar mass would be appropriate for the evaluation of the MMD, and another, separating by chemical composition, would be suited for determining the CH of the copolymer. In general, fractionation is influenced by the molar mass *and* chemical composition, and the direction of separation is determined by the experimental conditions, in particular by the solvent/non-solvent combination chosen.

4.2.1 Copolymer Analysis by Gradient HPLC

As was discussed in Sect. 2, chromatographic retention may be directed by entropic and enthalpic interactions. In particular, enthalpic interactions of the

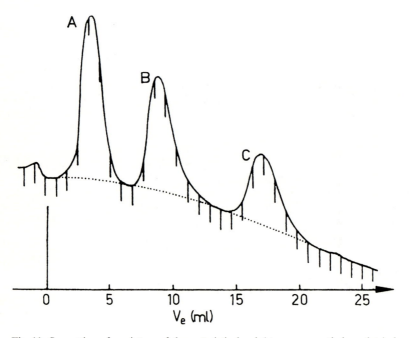

Fig. 11. Separation of a mixture of three statistical poly(styrene-*co*-methyl acrylate)s by gradient HPLC; stationary phase silica gel; mobile phase: carbon tetrachloride-methyl acetate sample composition: 42% (A), 52% (B), 74% (C) methyl acrylate. (From [34] with permission)

solute molecules and the stationary phase may be used for the separation of copolymers with respect to chemical composition. For stationary phases of a certain polarity, very specific precipitation/redissolution processes are able to promote separation with respect to chemical composition. Using solvent mixtures as the mobile phase, the precipitation/redissolution equilibria may be adjusted, and by changing the composition of the mobile phase during the elution process the solubility of the sample fractions may be changed. Thus, using gradient elution techniques, the polymer sample may be fractionated with respect to solubility and, accordingly, with respect to chemical composition.

One of the first separations of statistical copolymers using gradient HPLC was carried out by Teramachi et al. [34]. Mixtures of poly(styrene-*co*-methyl acrylate)s were separated by composition on silica columns through a carbon tetrachloride/methyl acetate gradient (see Fig. 11). When increasing the content of methyl acetate in the eluent, retention increased with increasing methyl acrylate content in the copolymer. This behavior fitted the normal-phase chromatographic system used. Similar separations could be achieved on other columns as well, such as polar bonded-phase columns (diol, nitrile, amino columns) [1].

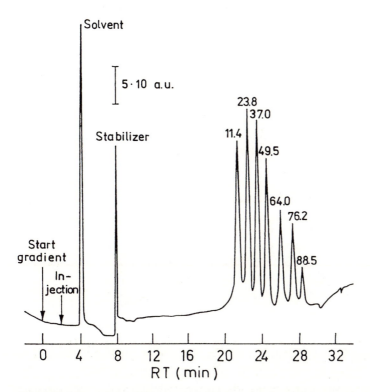

Fig. 12. Separation of a mixture of seven statistical poly(styrene-*co*-methyl methacrylate)s by gradient HPLC: stationary phase silica gel; mobile phase iso-octane-(THF + 10% MeOH); samples: mass % of MMA indicated. (From [35] with permission)

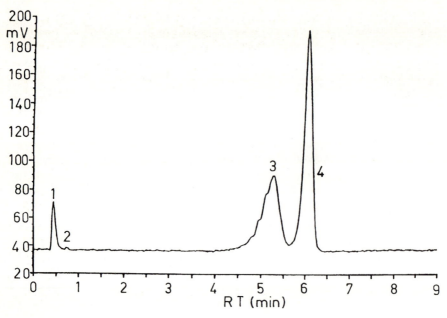

Fig. 13. Separation of EPDM-*graft*-PMMA by gradient HPLC: stationary phase cyano-propyl-silica gel; mobile phase iso-octane-THF. (From [36] with permission)

Copolymers of styrene and methyl methacrylate were separated by composition in numerous eluents. Most of them represented proper normal-phase systems with gradients increasing in polarity and a polar stationary phase. Figure 12 shows the separation of the mixture of seven statistical poly(styrene-*co*-methyl methacrylate)s on a silica column through a gradient i-octane/(THF + 10% methanol) [35].

Graft copolymerization usually yields mixtures containing the desired graft copolymer, non-grafted precursor molecules, and homopolymers formed as a by-product during the grafting reaction. Figure 13 represents the separation of a graft copolymer of methyl methacrylate onto EPDM by gradient HPLC in i-octane/THF on a cyanopropyl-modified silica column [36]. The first peak corresponds to the non-grafted EPDM, whereas peak 3 shows the desired graft copolymer. The more polar PMMA homopolymer was retained longer and eluted in peak 4.

4.2.2 Copolymer Analysis by Liquid Chromatography at the Critical Point of Adsorption

Another more recent approach to the chromatographic characterization of copolymers is the concept of "invisibility", which assumes that chromatographic conditions exist under which heteropolymers may be separated according to the size of one of the components because the other component is chromatographically "invisible", i.e. does not contribute to retention. Experimentally this concept relates to liquid chromatography at the critical point of adsorption, which was described in detail in Sect. 2. The usefulness of this technique for the determination of FTD of functional homopolymers was demonstrated in Sect. 4.1.

The application of liquid chromatography at the critical point of adsorption to block copolymers is based on the consideration that Gibbs free energy ΔG_{AB} of a block copolymer $A_n B_m$ is the sum of the contributions of block A and block B, ΔG_A and ΔG_B respectively:

$$\Delta G_{AB} = \Sigma n_A \Delta G_A + n_B \Delta G_B. \tag{12}$$

As specific interactions between blocks A and B are usually negligible in the chromatographic system, an interaction parameter χ_{AB} must not be introduced in Eq. (12).

Using chromatographic conditions, corresponding to the critical point of homopolymer A, block A in the block copolymer will be chromatographically invisible, and the block copolymer will be eluted solely with respect to block B:

$$\Delta G_A = 0$$

$$\Delta G_{AB} = n_B \Delta G_B \tag{13a–c}$$

$$K_d^{AB} = K_d^B.$$

Similarly, at the critical point of homopolymer B, block B will be chromatographically invisible, and the block copolymer will be eluted solely with respect to block A:

$$\Delta G_B = 0$$

$$\Delta G_{AB} = n_A \Delta G_A \tag{14a–c}$$

$$K_d^{AB} = K_d^A.$$

The first proof of the validity of this approach was given by Gankina et al. [19] for the analysis of block copolymers by thin layer chromatography. Column liquid chromatography was used by Zimina et al. [20] for the analysis of poly(styrene-*block*-methyl methacrylate) and poly(styrene-*block*-tert-butyl methacrylate). However, the critical conditions were established only for the polar part of the block copolymers, i.e. PMMA and P*t*BMA, respectively. Thus, only the polystyrene block was analyzed.

Depending on the polarity of blocks A and B in the block copolymer $A_n B_m$, and the polarity of the stationary phase, different chromatographic situations can be encountered (see Fig. 14). For example, at the critical point of homopolymer A, homopolymer B may be separated either in the SEC mode (1) or the adsorption mode (2). The same is true for the critical point of B, where A may be eluted according to (3) or (4). Preferable, of course, are cases (1) and (3),

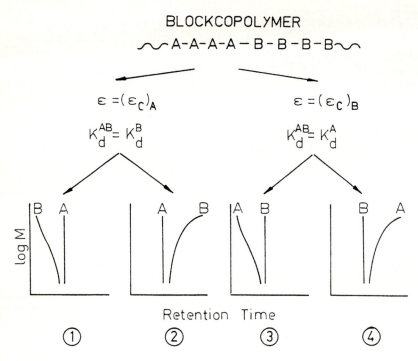

Fig. 14. Schematic representation of different chromatographic situations in liquid chromatography at the critical point of adsorption of block copolymers. For explanation of 1–4 see text

whereas, in cases (2) and (4) for high molar mass polymers, irreversible adsorption may be encountered. Let us now assume that the polarity of A is higher than the polarity of B. In this case, chromatographic behavior according to (1) is achieved when silica gel is used as the stationary phase (silica gel separates in order of *increasing* polarity). On the other hand, separation according to (3) is obtained on a reversed phase, such as RP-8 or RP-18 (separation in order of *decreasing* polarity).

After these initial considerations, the complete analysis of a number of diblock copolymers of styrene and methyl methacrylate shall be discussed in detail. The poly(styrene-*block*-methyl methacrylate)s under investigation were prepared via anionic polymerization of styrene and subsequent polymerization of methyl methacrylate, varying molar mass and composition (B1–B3). The polystyrene precursors (P1–P3) were isolated and characterized separately. As the PMMA block is the more polar block in the block copolymer, a polar (silica gel) column was chosen for establishing the critical point of PMMA. According to case (1) in Fig. 14, the PS block is then eluted in the SEC mode. The behavior of PMMA of different molar masses on silica gel Si-100 in eluents comprising methylethylketone and cyclohexane is shown in Fig. 15A [37].

At concentrations of methylethylketone > 73 vol.% a SEC mode is operating, whereas at concentrations < 73 vol.% of methylethylketone, adsorption takes place. The critical point of PMMA is obtained at an eluent composition of

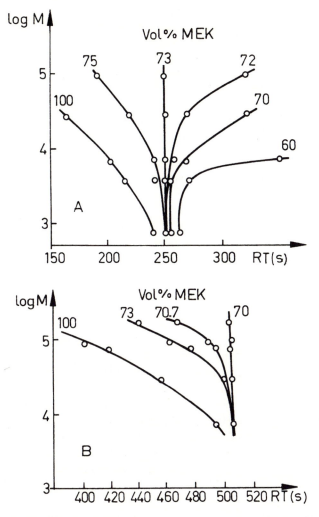

Fig. 15A, B. Critical diagrams molar mass vs retention time of PMMA: stationary phase: **A** Nucleosil Si-100; **B** LiChrospher Si-300 + Si-1000; mobile phase methylethylketone(MEK)-cyclohexane. (From [37] with permission)

methylethylketone-cyclohexane 73:27 v/v. At this point all PMMA samples are eluted in one peak. Accordingly, using these chromatographic conditions block copolymers containing PMMA as one block component may be analyzed with respect to the second block. Depending on the size of the macromolecules under investigation, like in conventional SEC the pore size of the stationary phase has to be adjusted to the desired molar mass region. Thus, for higher molar mass samples the investigations must be carried out on column sets with larger pores (see Fig. 15B) for a two-column set of Si-300 and Si-1000. For these columns the critical point of PMMA was obtained using methylethylketone-cyclohexane

70:30 v/v. The chromatographic behavior of the block copolymers and the corresponding polystyrene precursors in the SEC mode for both components (A) and in the critical mode for the PMMA block (B) are given in Fig. 16. The chromatograms clearly indicate that in (A) separation occurs with respect to the total molar mass of the samples, whereas in (B) the PMMA block is chromatographically invisible and separation is solely directed by the PS block, resulting in complete overlay of the chromatograms of B1 and P1, B2 and P2, and B3 and P3, where B1–B3 are the block copolymers and P1–P3 are the corresponding PS precursors. In agreement with the principles of liquid

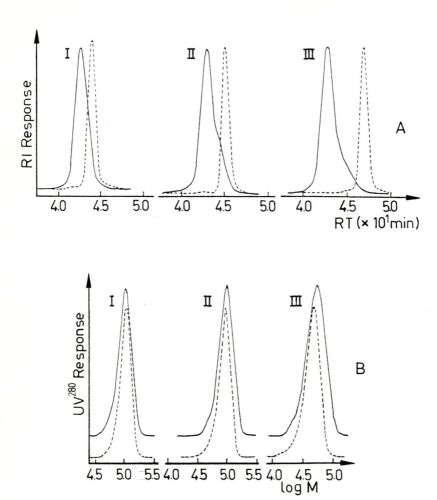

Fig. 16A, B. Chromatograms of poly(styrene-*block*-methyl methacrylate)s (———) and their polystyrene precursors (- - -): **A** in the SEC mode for both blocks: stationary phase styragel; mobile phase THF; **B** at the critical point for the PMMA block: stationary phase silica gel; mobile phase methylethylketone-cyclohexane 70:30 v/v: samples: I-B1,P1, II-B2,P2, III-B3,P3. (From [37] with permission)

chromatography at the critical point of adsorption, the analysis of the PMMA block has to be conducted on a different phase system. Formally, the critical point for PS can be established on different stationary phases. Using a silica gel column, at the critical point of PS the PMMA would be eluted in an adsorption mode – see case (4) in Fig. 14. As retention in the adsorption mode is exponentially related to molar mass, irreversible adsorption would be likely to occur for higher molar mass samples. Therefore, a reversed stationary phase was used, thus establishing conditions, corresponding to case (3) in Fig. 14. The retention behavior of PS on a column set of Nucleosil RP-18 with pore sizes of 300 Å and 1000 Å and an eluent of THF-acetonitrile is given in Fig. 17 [38]. The critical point of adsorption corresponds to an eluent composition of THF-acetonitrile 49:51 v/v. At the critical point of the PS block, the block copolymers are eluted exclusively with respect to the PMMA block, and using an appropriate PMMA calibration curve the molar mass and the polydispersity of this block in the block copolymers was calculated. Now, being able to determine the block lengths of the PS and the PMMA blocks separately, the total molar masses of the block copolymers and their chemical composition were determined from these data.

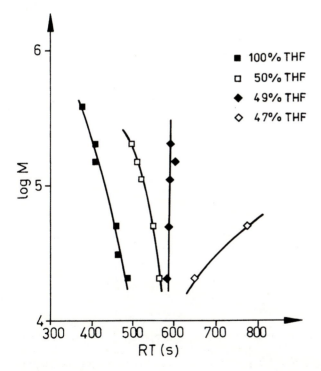

Fig. 17. Critical diagram molar mass vs retention time of polystyrene: stationary phase RP-18 300 Å + 1000 Å; mobile phase THF-acetonitrile. (From [38] with permission)

In order to compare the results of critical chromatography with results of an independent method, SEC with coupled density (D) and refractive index (RI) detection was used, which has been shown to be very useful for the characterization of copolymers with respect to their chemical composition [39, 40]. The MMD curve for one of the block copolymers and the mass distribution curves of the components are shown in Fig. 18. From these the overall chemical composition may be calculated. An excellent agreement between the results of critical chromatography and the SEC experiments was obtained.

According to Gorbunov and Skvortsov [18], triblock copolymers of the ABA type may be analyzed by liquid chromatography at the critical point of adsorption similar to the analysis of diblock copolymers. The two possible cases for this type of investigation, i.e. (a) the analysis with respect to the inner block B using the critical conditions of the outer block A, and (b) the analysis of the outer block A using the critical conditions of the inner block B, will be discussed briefly.

The separation of triblock copolymers of ethylene oxide (EO) and propylene oxide (PO), namely $HO(EO)_n(PO)_m(EO)_nOH$, at the critical point of the EO outer blocks is shown in Fig. 19 [16]. At this point the EO blocks were

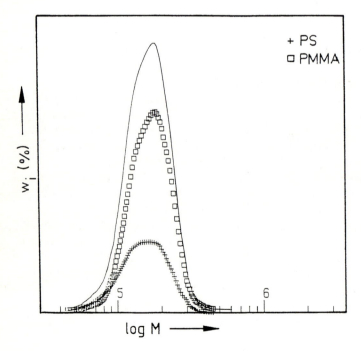

Fig. 18. MMD of the block copolymer B3 and and mass distributions of the components by SEC with coupled D-RI detection: stationary phase styragel; mobile phase chloroform. (From [38] with permission)

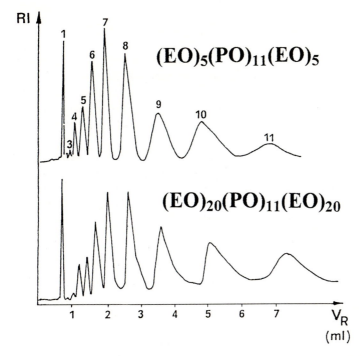

Fig. 19. Chromatograms of triblock copolymers of EO and PO at the critical point of EO: stationary phase RP-18; mobile phase acetonitrile-water 42:58 v/v. (From [16] with permission)

chromatographically invisible, and the block copolymers were eluted with respect to the chain length of the inner PO block. Different to the previous examples, the block to be analyzed was separated in the adsorption mode – see case (2) in Fig. 14. Accordingly, each peak in the chromatogram corresponded to one degree of polymerization with respect to PO. In this first separation step fractions were obtained which were uniform with respect to PO but had an MMD with respect to EO. Due to problems of solubility and specific interactions with the stationary phase it was not possible to determine the outer EO blocks at the critical point of PPO [16]. This could be achieved, however, by subjecting the PO uniform fractions to a second chromatographic method. This method must separate the fractions with respect to the oligomer distribution of the EO blocks, thus providing the MMD of these blocks. Fig. 20 shows the separation of the fractions by capillary SFC [41], where a separation into the EO oligomers is obtained. The concentration of each oligomer was determined, thus providing the MMD of the EO blocks. The same information was obtained when the fractions were investigated by SEC instead of capillary SFC [41].

The analysis of a triblock copolymer according to its outer blocks was carried out by using chromatographic conditions, substantiating the critical mode for the inner block. The behavior of a triblock copolymer of THF and

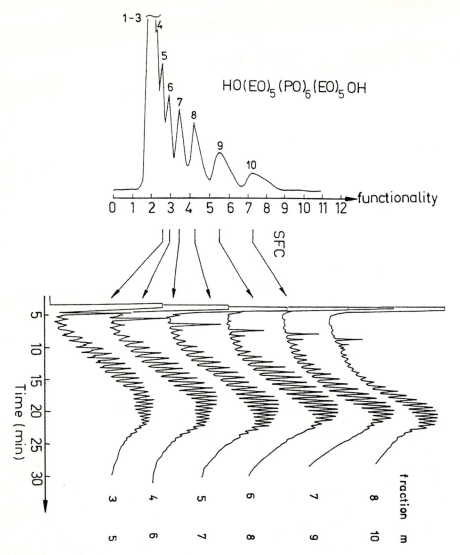

Fig. 20. Two-dimensional separation of a triblock copolymer of EO and PO, critical chromatography vs supercritical fluid chromatography. (From [41] with permission)

1,3,6-trioxocane is shown in Fig. 21. Using non-aqueous reversed phase chromatography on an RP-18 stationary phase the critical conditions were established for the inner poly-THF block. At this point the trioxocane outer blocks were eluted in the SEC mode, and the elution curves represent the molar mass distribution [42]. Via an appropriate SEC calibration procedure, the numerical data of MMD of the outer blocks were calculated.

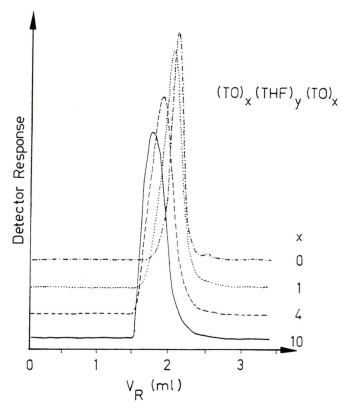

Fig. 21. Chromatograms of triblock copolymers of THF and 1,3,6-trioxocane(TO) at the critical point of poly-THF: stationary phase RP-18; mobile phase acetone-hexane 95:5 v/v. (From [42] with permission)

4.3 Analysis of Polymer Blends

Polymer blends, i.e. mixtures of two or more polymeric components, are of increasing commercial importance for a number of applications. The advantage of polymer blends is the useful combination of the properties of the components without creating chemically new polymers. This approach in many cases is more feasible than developing new tailor-made polymer structures.

In most cases the identification and quantitative determination of blend components is complicated and, depending on the chemical structure, a variety of different analytical methods must be used. Spectroscopic methods, such as infrared spectroscopy [43–45] and nuclear magnetic resonance [46–48] may help to identify blend components. For the determination of the MMD of the components, however, a separation step is often required.

Size exclusion chromatography has been used for the analysis of polymer blends by a number of authors [49–51]. However, as SEC separates according to hydrodynamic volume of the macromolecules, this method is limited to blends containing components of different molar mass. As already discussed, the separation of polymers according to differences in chemical composition and structure can be achieved by gradient elution HPLC (see Sect. 4.2). Using this technique, statistical copolymers were separated according to composition, and graft copolymers were separated into copolymer and homopolymer fractions. The separation of mixtures of poly(meth)acrylates by gradient HPLC has been successfully conducted by Mourey [52].

The effect of the alkoxy group on the retention behavior of the poly(meth)acrylates provided the chance of separating different (meth)acrylate homopolymers through normal-phase gradient elution. Fig. 22A shows the separation of polymethacrylate esters on silica gel using a toluene-methylethylketone gradient. As was expected from the polarity of the alkoxy group, polybutyl methacrylate was retained less than polymethyl methacrylate. The same sequence holds for polyacrylates (see Fig. 22B). From the chromatograms the composition of the samples may be quantitatively determined. As in this case separation is accomplished with respect to chemical composition and not molar mass, a second chromatographic method must be used for the determination of the MMD of the chromatographic fractions.

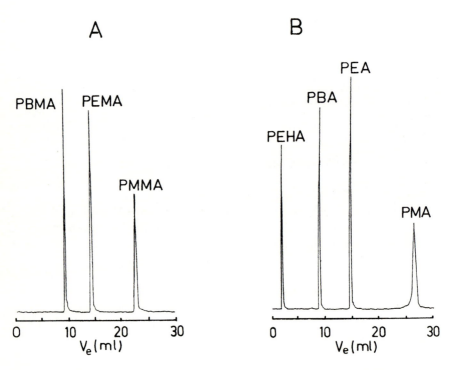

Fig. 22A, B. Separation of: **A** polymethacrylates; **B** polyacrylates by gradient HPLC: stationary phase silica gel; mobile phase toluene-methylethylketone. (From [52] with permission)

The general behavior of a binary blend in different chromatographic modes is summarized in Fig. 23 [53]. In the size exclusion and the adsorption modes for both blend components, their retention behavior is very similar, and the calibration curves log molar mass vs retention time suggest that, in these cases, one retention time normally corresponds to two molar masses (one molar mass on each calibration curve). Thus, for comparable molar masses of the components, an overlapping of the elution zones is obtained. Accordingly, sufficient separation of the components using SEC or adsorption chromatography may be achieved only when their molar masses are quiet different.

A completely different behavior of the blend components is obtained when chromatographic conditions are used, corresponding to the critical mode of one of the components. In this case, the elution zones are separated from each other over the entire molar mass range, and separation is achieved even for components of similar molar mass.

To explain the separation procedure of polymer blends using chromatography at the critical point of adsorption, the behavior of blends of polystyrene (PS) and polymethyl methacrylate (PMMA) in different chromatographic modes is shown in Fig. 24. With silica gel Si-100 as the stationary phase, the mobile phase comprised mixtures of MEK and cyclohexane. In pure MEK a size exclusion mode was operating for both components. Under these conditions PS and

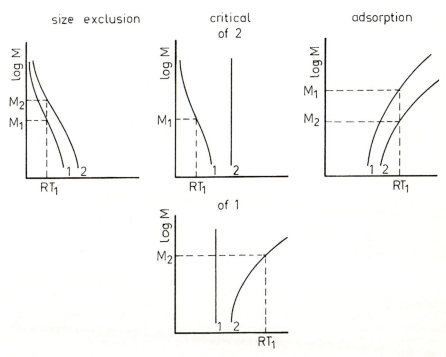

Fig. 23. Behavior of the calibration curves log molar mass vs retention time of a binary polymer blend in the three chromatographic modes. (From [53] with permission)

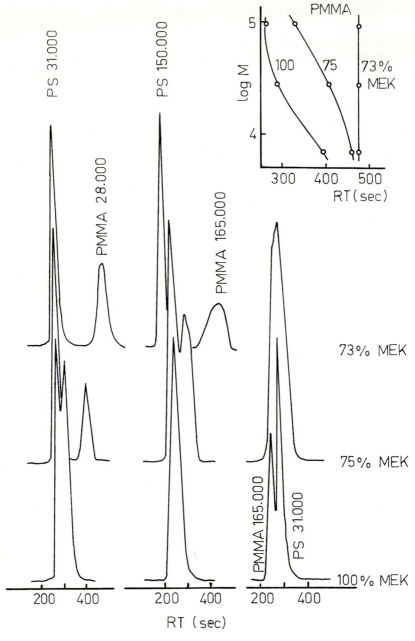

Fig. 24. Chromatograms of PS-PMMA blends in different chromatographic modes, and critical diagram log molar mass vs retention time for PMMA (*upper right corner*): stationary phase silica gel Si-100; mobile phase methylethylketone-cyclohexane. (From [53] with permission)

PMMA may be separated only if their molar masses are different. For low molar mass samples (PS, PMMA $\sim$ 30 000 g/mol) the two components of the blend were identified, but separation was poor. For higher molar masses (PS, PMMA $\sim$ 150 000 g/mol), however, one symmetric elution peak, similar to the elution profile of a homopolymer, was obtained. When cyclohexane was added to MEK, the elution behavior of PMMA changed dramatically, whereas for PS it remained nearly constant. In the near-critical region (75 vol.% of MEK) separation of blend components of equal molar mass was improved, compared to the size exclusion mode. At the critical point of PMMA, which corresponds to a mobile phase composition of MEK-cyclohexane 73:27 vol.% (see Fig. 24, upper right corner), a complete separation of the elution zones of PMMA and PS was achieved. Regardless of the molar mass, all PMMA fractions were eluted

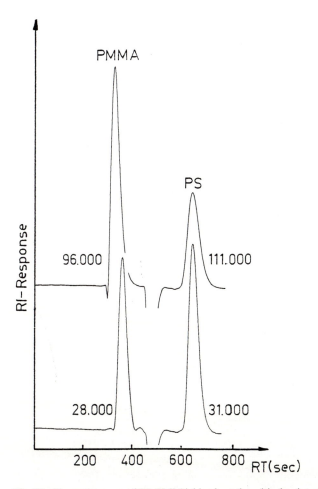

Fig. 25. Chromatograms of PS-PMMA blends at the critical point of PS: stationary phase RP-18; mobile phase THF-water 88.8:11.2 v/v

at the same retention time, whereas for the PS fractions a size exclusion mode was operating, and retention time decreased with increasing molar mass. Accordingly, the molar mass and the polydispersity of the PS blend components may be determined via a conventional SEC calibration curve.

In a similar approach, the molar mass distribution of the PMMA fraction in the blends may be determined when chromatographic conditions are operating corresponding to the critical mode of PS and the size exclusion mode of PMMA.

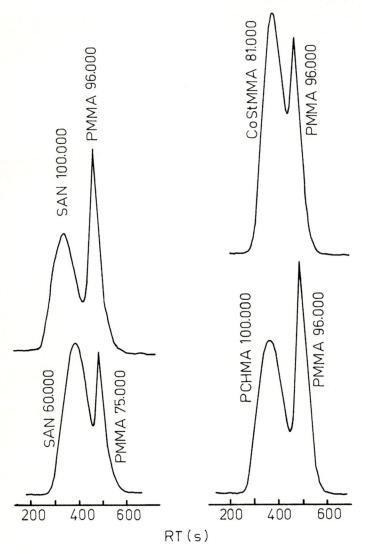

Fig. 26. Chromatograms of blends of PMMA and copolymers at the critical point of PMMA: stationary phase silica gel Si-300 + Si-1000; mobile phase methylethylketone-cyclohexane 70:30 v/v. (From [53] with permission)

This was achieved on a reversed stationary phase RP-18 with a mobile phase of THF-water 88.8:11.2 v/v (see Fig. 25).

Since, at the critical point of a certain polymer, this polymer is always eluted at the same retention time (corresponding to $K_d = 1$), all different types of polymer blends containing this polymer as one component may be separated. This is demonstrated in Fig. 26 for PMMA- containing blends. Even chemically very similar blend components, such as poly(cyclohexyl methacrylate) (CHMA) and PMMA were separated. In addition, not only blends of homopolymers, but also blends of copolymers and PMMA may be investigated. Like the homopolymers, poly(styrene-*co*-acrylonitrile) and poly(styrene-*co*-methyl methacrylate) (CoStMMA) were eluted in the SEC mode.

5 Applications and Future Developments

5.1 Applications

The following Tables 2–4 briefly summarize some recent work on the application of interaction chromatography to the analysis of complex polymers. All major

Table 2. Analysis of functional homopolymers by interaction chromatography

Polymer	Stationary Phase	Eluent	Reference
oligocarbonates	silica gel	chloroform/carbontetrachloride	15
oligosulfones	silica gel	chloroform/carbontetrachloride	54
polypropylene glycol	silica gel	methylethylketone/ethylacetate	11, 14
polyethyleneglycol adipate	silica gel	methylethylketone/chloroform	11
polypropyleneglycol adipate	silica gel	methylethylketone/chloroform	11
epoxy resin	silica gel	not given	12, 13
polyethylene oxide	RP-18	methanol/water	55, 66
		acetonitrile/water	16
polytrioxocane	RP-18	acetonitrile/water	56–59
polybutyleneglycol terephthalate	silica gel	tetrahydrofurane/heptane	60
polybutyleneglycol adipate	silica gel	acetone/hexane	61
amino-terminated polypropylene oxide	RP-18, RP-8	methanol/water	62
		acetonitrile/water	62
aliphatic polyester	silica gel	acetone/hexane	63
alkylphenol-PEO	silica gel	hexane + ether/ethanol + water	64
	RP-18	methanol/water	64
polybutylene glycol	RP-18, RP-4	acetonitrile/water	65
		methanol/water	65

Table 3. Analysis of copolymers by interaction chromatography

Polymer	Stationary Phase	Eluent	Reference
poly(St-co-MMA)	crosslinked PAN	chloroform/hexane	67
	silica gel	tetrahydrofurane/hexane	67
poly(St-co-MA)	modified silica gel	heptane/(dichloromethane + methanol)	68
poly(St-b-MMA)	silica gel	trichloromethane/ethanol	69
poly(St-b-t-BMA)	silica gel	i-octane/tetrahydrofurane	70
PMMA (it, st)	crosslinked PAN	hexane/dichloromethane	71
poly(St-co-t-BMA)	RP-18	methanol/tetrahydrofurane	72
poly(DMA-b-MMA)	cyanopropyl silica gel	i-octane/tetrahydrofurane	73
poly(St-co-2-methoxy-methyl methacrylate)	cyanopropyl silica gel	i-octane/tetrahydrofurane	74
poly(St-co-MA)	silica gel	chloroform/ethanol	75, 76
poly(St-b-VA)	silica gel	dichloroethane/ethanol	77
poly(EMA-co-BMA)	silica gel	dichloroethane/ethanol	78
poly(St-co-EMA)	silica gel	dichloroethane/ethanol	79
poly(EPDM-g-MMA)	cyanopropyl silica gel	i-octane/tetrahydrofurane	36
poly(MMA-co-MAS)	silica gel	methanol/toluene	80
poly(DMA-b-MMA)	silica gel	methylethylketone/cyclohexane	81
	RP-18	ethanol/chloroform	82
poly(St-b-MMA)	silica gel	methylethylketone/cyclohexane	37
	RP-18	tetrahydrofurane/acetonitrile	38
poly(EO-b-PO-b-EO)	RP-18	acetonitrile/water	41
(St-Bd) stars	silica gel	i-octane/THF	83
poly(MAS-co-MMA)	silica gel	methanol/toluene	84
PMMA-g-polydi-methylsiloxane	silica gel	ethanol/toluene/cyclohexane	85

Table 4. Analysis of polymer blends by interaction chromatography

Blends	Stationary Phase	Eluent	Reference
PS-PPE	cyanopropyl silica gel	i-octane/THF/chloroform	86
PS-EPD	cyanopropyl silica gel	methanol/toluene	86
PVC-PMMA	silica gel	methylethylketone/cyclohexane	53
PS-PMMA	silica gel	dichloroethane/chloroform	87
PS-PnBMA	silica gel	heptane/tetrahydrofurane	51
PS-PMMA	silica gel	methylethylketone/cyclohexane	53
SAN-PMMA	silica gel	methylethylketone/cyclohexane	53
poly(St-co-MMA)-PMMA	silica gel	methylethylketone/cyclohexane	53

work on the determination of FTD of functional homopolymers before 1986 is compiled in the review of Entelis et al. [2], whereas in Glöckner's book all work regarding the analysis of copolymers before 1989 is summarized [1]. Therefore, in the following tables only work is included, which appeared after 1986 and 1989, respectively.

5.2 Conclusions and Future Developments

Heterogeneous or complex polymers are distributed in more than one molecular parameter. For functional homopolymers one has to deal with the overlapping effects of molar mass distribution and functionality type distribution, whereas copolymers are distributed at least in molar mass and chemical composition. For many years, detector development and the use of several detectors attached to SEC have been the major thrusts in chromatographic analysis of complex macromolecules. In particular, the combination of a refractive index and an ultraviolet detector has been used extensively, although only a limited number of polymers is UV active. Therefore the application of this technique is certainly not universal. On the other hand, SEC has its merits in the daily routine because it is simple, fast, and very reproducible.

The development of gradient HPLC of polymers was a first major step towards separating copolymers according to their chemical composition. By combining gradient HPLC and SEC in the chromatographic cross-fractionation technique, a two-dimensional information molar mass vs chemical composition of copolymers was obtained for the first time solely by column chromatography. However, the major drawback of the wellknown SEC and gradient HPLC techniques is that they do not separate according to one heterogeneity alone. SEC separates according to hydrodynamic volume, which for copolymers is a function of molar mass *and* chemical composition and, therefore, is not a direct measure of MMD. On the other hand, gradient HPLC as a mixed-mode chromatographic technique separates macromolecules according to polarity, hydrophobic properties or solubility, which again are influenced by both MMD and chemical composition. Thus, when attempting to separate with respect to chemical composition, in the general case the molar mass influence must always be taken care of.

To summarize, a chromatographic separation technique is desirable which separates according to only one heterogeneity, for example FTD, making sure that molecular size is not an influencing parameter.

The most promising approach to the separation of complex polymers with respect to only one heterogeneity is liquid chromatography at the critical point of adsorption. In this technique, the ratio of entropic and enthalpic effects in chromatography are influenced on purpose by varying the chromatographic conditions, such as polarity of the stationary phase, composition of the eluent, temperature etc. Thus, by balancing the entropic and enthalpic effects, a chromatographic mode is established, where the size of the polymer chain does not affect retention. In this case retention is governed solely by the heterogeneities of the polymer chain. Using liquid chromatography at the critical point of adsorption, functional homopolymers and block copolymers are separated with respect to FTD and the structure of the single blocks, respectively.

As heterogeneous polymers are distributed in more than one molecular parameter, more than one chromatographic separation technique must be used. For functional homopolymers evidence is first obtained that the optimum separation protocol includes liquid chromatography at the critical point of adsorption as the first dimension of separation, yielding fractions which are homogeneous in functionality. When these fractions are subjected to any molar mass sensitive separation technique, MMD for each functionality fraction, and therefore the complete FTD-MMD relationship, is obtained. Two-dimensional separations of this type are very much susceptible to automation, as has been shown by Much et al. [88] and Kilz and coworkers [89–91].

For the analysis of block or graft copolymers, liquid chromatography at the critical point of adsorption may be used to determine the single blocks (see Sect. 4.2). Making one of the blocks chromatographically invisible, the other block(s) may be analyzed. Thus for diblock copolymers, at the critical point of block A, block B is analyzed, and vice versa [92]. As separation is very sensitive to the eluent composition, the on-line combination of two critical chromatography modes is not useful. Much more feasible from the point of view of automation is the separation of block copolymers in the critical mode for block A (corresponding to the SEC mode for block B) in the first dimension, and the analysis of the fractions using SEC in the second dimension. As in the first separation step fractions are obtained which are monodisperse with respect to block B and polydisperse, with respect to block A, the second dimension unambiguously yields the MMD of block A. Thus, combining critical chromatography and SEC, block copolymers can be fully described with respect to their individual blocks.

Once sophisticated separation techniques are available, one can think about optimizing detection. Apart from the classical combination of RI and UV detectors, the coupling of two really universal detectors holds some promise. In particular, the combination of an RI and a density detector seems to be a good choice for a number of applications [40, 93, 94]. In general, any polymer generates a signal in the RI and the density detector, and the sensitivities of different polymers in both detectors are different. Accordingly, via dual RI-density detection the composition of heterogeneous polymers can be determined at each point of an elution curve. A significant limitation of this technique is, however, that it may be used only in isocratic separation modes, because the eluent composition affects the detector response.

With respect to specific detectors, the development of a new interface for FTIR detection must be mentioned. This interface, which is coupled to the HPLC system, is a solvent removal device which provides samples in a solvent free condition. Unlike flow-through cells, where the eluent absorption peaks interfere with those of the sample, in the new device the eluent is removed before

FTIR measurement. Thus the spectral characteristics of the sample are exclusively obtained [95–98].

In conclusion, mass spectrometry has recently been shown to be an interesting alternative to the existing detection systems. Matrix-assisted laser desorption/ionization mass spectrometry (MALDI-MS), developed by Karas and Hillenkamp [99], has been used successfully to determine the masses of large biomolecules and synthetic polymers [100]. The advantage of MALDI-MS over other MS techniques is the soft ionization of the sample molecules, and the significantly reduced fragmentation. As a result, the sample is separated according to the mass of its components and for heterogeneous polymers a molar mass distribution may be obtained. As minimum quantities of a few nanograms are sufficient for a proper analysis, chromatographic fractions can be analyzed very efficiently. It has been shown that one drop of the eluate after an analytical separation is adequate for the MALDI-MS experiment [32, 101].

6 References

1. Glöckner G (1991) Gradient HPLC of copolymers and chromatographic cross-fractionation. Springer, Berlin Heidelberg New York
2. Entelis SG, Evreinov VV, Gorshkov AV (1986) Adv Polym Sci 76: 129
3. Balke ST (1991). In: Barth HG, Mays JM (eds) Modern methods of polymer characterization. John Wiley, New York
4. Glöckner G (1987) Polymer characterization by liquid chromatography. Elsevier, Amsterdam
5. Balke ST (1984) Quantitative column liquid chromatography. A survey of chemometric methods. Elsevier, Amsterdam
6. Glöckner G (1986) Adv Polym Sci 79: 159
7. Entelis SG, Evreinov VV, Kuzaev AI (1985) Reactive Oligomers (Russ.). Khimiya, Moscow
8. Tennikov MB, Nefedov PP, Lazareva MA, Frenkel SJ (1977) Vysokomol Soedin A19: 657
9. Belenkii BG, Gankina ES, Tennikov MB, Vilenchik LZ (1976) Dokl Acad Nauk USSR 231: 1147
10. Skvortsov AM, Belenkii BG, Gankina ES, Tennikov MB (1978) Vysokomol Soedin A20: 678
11. Filatova NN, Gorshkov AV, Evreinov VV, Entelis SG (1988) Vysokomol Soedin A30: 953
12. Gorshkov AV, Verenich SS, Evreinov VV, Entelis SG (1988) Chromatographia 26: 338
13. Gorshkov AV, Verenich SS, Markevich MA, Pektinov VI, Evreinov VV, Entelis SG (1989) Vysokomol Soedin A31: 1878
14. Gorshkov AV, Evreinov VV, Lausecker B, Pasch H, Becker H, Wagner G (1986) Acta Polymerica 37: 740
15. Gorshkov AV, Prudskova TN, Guryakova VV, Evreinov VV (1986) Polym Bull 15: 465
16. Gorshkov AV, Much H, Becker H, Pasch H, Evreinov VV, Entelis SG (1990) J Chromatogr 523: 91
17. Gorbunov AA, Skvortsov AM (1988) Vysokomol Soedin A30: 453
18. Gorbunov AA, Skvortsov AM (1988) Vysokomol Soedin A30: 895
19. Gankina E, Belenkii B, Malakhova I, Melenevskaya E, Zgonnik VJ (1991) Planar Chromatogr 4: 199
20. Zimina TM, Kever JJ, Melenevskaya EY, Fell AF (1992) J Chromatogr 593: 233
21. Abbott SR, Tusa J (1983) J Liquid Chromatogr 6: 77
22. Scott RPW (1986) Liquid Chromatography Detectors. Elsevier, Amsterdam
23. Yeung ES (1986) Detectors for Liquid Chromatography. John Wiley & Sons, New York
24. Shafer KH, Pentoney SL, Griffiths PR (1984) J High Res Chromatogr, Chromatogr Commun 7: 707

25. Willis JM, Dwyer JL, Wheeler L (1993) International Symposium on Polymer Analysis and Characterization. Crete, Greece July 11
26. Hamielec AE, Ouano AC, Nebenzahl LL (1978) J Liquid Chromatogr 1: 527
27. Haney MA (1985) J Appl Polym Sci 30: 3037
28. Yau WW, Abboutt SD, Smith GA, Keating MY (1987) Am Chem Soc Symp Ser 352: 80
29. Muenker AH, Hudson BE (1969) J Macromol Sci A3: 1465
30. Evreinov VV (1970) Vysokomol Soedin A12: 829
31. Gorshkov AV (1985) Vysokomol Soedin B27: 181
32. Pasch H, Zammert I (1994) J Liquid Chromatogr 17: 3091
33. Pasch H unpublished
34. Teramachi S, Hasegawa A, Shima Y, Akatsuka M, Nakajima M (1979) Macromolecules 12: 992
35. Glöckner G, van den Berg JHM (1986) J Chromatogr 352: 511
36. Augenstein M, Stickler M (1990) Makromol Chem 191: 415
37. Pasch H, Brinkmann C, Gallot Y (1993) Polymer 34: 4100
38. Pasch H, Gallot Y, Trathnigg B (1993) Polymer 34: 4986
39. Trathnigg B (1991) In: Lemstra PJ, Kleintjens LA (eds) Integr Fundam Polym Sci Technol. Proc Int Meet Polym Sci Technol Vol 5 Elsevier Applied Science, London
40. Trathnigg B, Yan X (1992) Chromatographia 33: 467
41. Pasch H, Brinkmann C, Much H, Just U (1992) J Chromatogr 623: 315
42. Pasch H, Much H, Schulz G, Gorshkov AV (1992) LC-GC International 5: 38
43. Gregoriou VS, Noda I, Dowrey AI, Marcott C, Chao JL, Palmer PA (1993) J Polym Sci. Polym Phys 31: 1769
44. Nishioka T, Teramae N (1993) J Appl Polym Sci. Polym Symp 52: 251
45. Ghebremeskel Y, Fields J, Garton A (1994) J Polym Sci. Polym Phys 32: 383
46. Schmidt P, Dybal J, Straka J, Schneider B (1993) Makromol Chem 194: 1757
47. Kelts LW, Landry CJ, Teengarden DM (1993) Macromolecules 26: 2941
48. Huo PP, Cebe P (1993) Macromolecules 26: 5561
49. Cantow HJ, Probst J, Stojanov C (1968) Kautschuk u Gummi 21: 609
50. Schröder E, Franz J, Hagen E (1976) Ausgewählte Methoden der Plastanalytik. Akademie-Verlag, Berlin
51. Balke ST (1982) Sep Purif Methods 11: 1
52. Mourey TH (1986) J Chromatogr 357: 101
53. Pasch H (1993) Polymer 34: 4095
54. Guryakova VV, Pavlov AV (1986) J Chromatogr 365: 197
55. Winkle W (1990) Chromatographia 29: 530
56. Krüger H, Much H, Schulz G, Wehrstedt C (1990) Makromol Chem 191: 907
57. Krüger H, Pasch H, Much H, Gancheva V, Velichkova R (1992) Makromol Chem 193: 1975
58. Pasch H, Krüger H, Much H, Just U (1992) J Chromatogr 589: 295
59. Pasch H, Krüger H, Much H, Just U (1992) Polymer 33: 3889
60. Prudskova TN, Guryakova VV, Pavlov AV (1992) Vysokomol Soedin A43: 148
61. Much H, Krüger RP, Schulz G (1993) Instrument Anal Chem and Computer Technol. Düsseldorf, Germany
62. Rissler K (1994) J Chromatogr 667: 167
63. Krüger R.-P., Much H, Schulz G (1994) J Liquid Chromatogr 17: 3069
64. Anghel DF, Balcan M, Voicu A, Elian M (1994) J Chromatogr 668: 375
65. Rissler K, Fuchslueger U (1994) J Liquid Chromatogr 17: 2791
66. Meyer T, Harms D, Gmehling J (1993) J Chromatogr 645: 135
67. Sato H, Takeuchi H, Tanaka Y (1986) Macromolecules 19: 2613
68. Sparidans RW, Claessens HA, van Doremaele GHJ, van Herk AM (1990) J Chromatogr 508: 319
69. Mori S (1989) J Appl Polym Sci 38: 95
70. Glöckner G, Müller AHE (1989) J Appl Polym Sci 38: 1761
71. Sato H, Sasaki M, Ogino K (1989) Polym J 21: 965
72. Glöckner G, Wolf D (1992) Chromatographia 34: 363
73. Augenstein M, Müller MA (1990) Makromol Chem 191: 2151
74. Glöckner G, Stickler M, Wunderlich W (1989) J Appl Polym Sci 37: 3147
75. Mori S, Mouri M (1989) Anal Chem 61: 2171
76. Mori S (1990) J Chromatogr 507: 473
77. Mori S (1990) J Chromatogr 503: 411
78. Mori S (1990) Anal Chem 62: 1902

79. Mori S (1991) J Chromatogr 541: 375
80. Schunk TC (1994) J Chromatogr 661: 215
81. Pasch H, Augenstein M (1993) Makromol Chem 194: 2533
82. Pasch H, Augenstein M, Trathnigg B (1994) Macromol Chem Phys 195: 743
83. Kilz P, Krüger R.-P., Much H, Schulz G (1995) In: Chromatographic characterization of polymers (Ed Provder T) ACS Adv Polym Ser 247: 223
84. Schunk TC (1994) J Chromatogr 661: 215
85. Schunk TC, Long TE (1995) J Chromatogr 692: 221
86. Jansen JAJ, van den Bungelaar JHJ, Leenen AJH (1991) In: Lemstra PJ, Kleintjens LA (eds) Integr Fundam Polym Sci Technol. Proc Int Meet Polym Sci Technol. Vol 5, Elsevier Applied Science, London
87. Mori S (1987) J Chromatogr 411: 355
88. Schulz G, Krüger RP, Much H (1994) International GPC Symposium '94. Lake Buena Vista, FL, June 7
89. Kilz P (1992) Labor Praxis 16: 628
90. Kilz P, Krüger R.-P., Much H, Schulz G (1993) PMSE Preprints 69: 114
91. Kilz P, (1995) ACS Preprints Analyt Soc. Anaheim, USA
92. Pasch H (1993) GIT Fachz Lab 37: 1068
93. Trathnigg B (1991) GIT Fachz Lab 35: 35
94. Trathnigg B, Maier B, Thamer D (1994) J Liquid Chromatgr 17: 4285
95. Willis JM, Dwyer JL, Liu MX (1994) International GPC Symposium '94, Lake Buena Vista, FL, June 7
96. Company information Lab Connections, 5 Mount Royal Ave, Marlborough, MA, USA
97. Cheung P, Balke ST, Schunk TC, Mourey TH (1993) J Appl Polym Sci. Appl Polym Symp 52: 105
98. Schunk TC, Balke ST, Cheung P (1994) J Chromatogr 661: 227
99. Karas M, Hillenkamp F (1988) Anal Chem 60: 2299
100. Beavis RC, Chait BT (1989) Rapid Commun Mass Spectrom 3: 436
101. Pasch H, Rode K (1995) J Chromatogr 699: 21

Editor: H. Beniot and H. Höcker
Received: May 1996

Wavelength Sensitivity in Polymer Photodegradation

Anthony L. Andrady
Camille Dreyfus Laboratory, Research Triangle Institute, Research Triangle Park, NC 27709, USA

This review examines the terminology and experimental approaches used to describe wavelength sensitivity in polymer photodegradation and summarizes the data published on the subject. Wavelength sensitivity or the efficacy of different regions or individual wavelengths in the ultra violet-visible region of the spectrum is usually expressed in terms of action and activation spectra. An action spectrum shows the effectiveness of a photoprocess at unit or constant irradiating photons, at different irradiating wavelengths. An activation spectrum shows the efficacy of different spectral regions of a specific source spectrum in causing a photoprocess. Published data on both naturally-occurring polymers and the key synthetic polymers are summarized.

Advances in Polymer Science, Vol. 128
© Springer-Verlag Berlin Heidelberg 1997

1 Introduction

The potential of a polymer for light-induced degradation is determined by its ability to absorb photons of suitable energy and the availability of photochemical pathways to utilize the absorbed energy for chemical reactions. The Grotthus-Draper Law states that only the radiation absorbed by a system is effective in producing chemical changes. Most polymers can absorb ultraviolet (UV) radiation of $\lambda < 300$ nm, while those with chromophores such as carbonyl groups and unsaturated centers can absorb even longer wavelengths of UV radiation. Pure aliphatic polymers are, however, transparent to solar radiation, but the presence of impurities and additives often allows them to absorb sunlight and undergo photodegradation. The photon energies associated with the near ultraviolet (UV) and visible radiation are in the same range as the bond dissociation energies of common covalent bands in organic molecules including polymers. With polymers, the main-chain bond dissociation energy is essentially the difference between the activation energies for initiation and termination. The unsubstituted C–C main chain bond and the C–H bond abundantly found in common polymers have bond energies in the range of 60–90 kcal per mole, and 80–100 kcal per mole, respectively. Figure 1 makes this comparison between the energy needed to dissociate a main chain bond in a polymer into macroradicals [1] and the corresponding wavelength of ultraviolet (UV) or visible radiation. Light-induced degradation, however, rarely proceeds via direct photolysis of the main-chain covalent linkages of the polymer; often a complex sequence of reactions involving excitation, energy transfer, and even oxidation is involved. Therefore, absorption of UV-visible radiation need not always lead to degradation; also, photodegradative main chain scission can occur as a result of the absorption of radiation of energy lower than the main chain bond dissociation energy of the polymer.

The primary consequence of the absorption of radiation by a macromolecule is the occurrence of photophysical processes [3]. Photon energy excites organic molecules, promoting spin-paired electrons in the molecular orbitals in the ground state molecules into higher energy levels. This excitation, when achieved with conservation of opposing spin-angular momentum of the electrons, yields excited singlet state, S_n, species. The energy in each excited state is quantized, and the species have a finite lifetime. Promotion of electrons to the lowest singlet energy state is usually the immediate result of the absorption of a photon; where higher energy singlet states are achieved they rapidly decay to the singlet state with lowest energy. As the electrons in the excited singlet state do not share the same orbital, they are without violating the Pauli exclusion principle, able to adopt parallel spins yielding excited triplet states. This singlet to triplet transitions by spin inversion must occur via intersystem crossing as a direct triplet excitation is a forbidden transition. The triplet states have lower energies than the corresponding singlet states, making transitions such as $S_1 \rightarrow T_1$ feasible. In addition to electronic excitation, absorption of electromagnetic

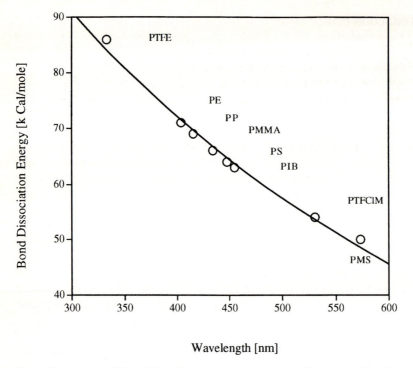

Fig. 1. Comparison of bond dissociation energies of main chain covalent bonds in common polymers to the wavelength of UV-Visible radiation of comparable energy. The line shows the logarithm of photon energy as a function of the wavelength

radiation may also cause transitions which alter the vibrational and rotational energy levels of a molecule. Each photon absorbed by a molecule may therefore potentially promote an electron into one of many vibrational levels within one of several available excited states. The absorption spectra of organic molecules therefore often tend to be complex and consist of broad bands (as opposed to line spectra obtained with atomic species).

The fate of these excited molecules is conveniently depicted using a Jablonskii diagram, showing the various radiative and nonradiative routes available for dissipation of the absorbed radiant energy. These include several important radiative photophysical processes such as fluorescence (reversion of $S_1 \rightarrow S_0$), phosphorescence (reversion of $T_1 \rightarrow S_0$), as well as nonradiative transitions. Radiative processes result in the emission of one or more wavelengths of radiation as excited-state electrons with differing energies participate in these processes. Photochemical processes essentially compete with these different physical mechanisms of energy dissipation. Due to their longer lifetimes of about 10^5 times that of the corresponding singlet state molecules, photochemical reactions generally tend to involve excited triplet state (mostly the lowest energy, T_1, excited states). These may undergo collisional energy transfers or associate with other species to form excimers or exciplexes through bimolecular collisions. Excited state species can be quenched by molecular oxygen as well [4], but the process can be diffusion-controlled in solid polymers [5]. Excited species will invariably dissipate their energy through either the very efficient photophysical processes or through relatively inefficient photochemical processes.

When photochemical pathways are available for the purpose, they involve sequences of chemical reactions with specific energy requirements. The number of possible photochemical degradative reactions of polymers is quite large and their detailed discussion is beyond the scope of this review. Excellent reviews on the subject are available [6–11]. As the photon energy increases with decreasing wavelength of radiation it is reasonable to expect the high-energy short-wavelength radiation to be more effective in promoting a wider range of these reactions. Consequently, polymers exhibit marked wavelength sensitivity in their photodegradation behavior. This is a consequence of the differences in energy of absorbed photons of different wavelengths and also due to the absorption characteristics of polymers (all wavelengths not being equally absorbed by the polymer). Most of the interest in wavelength sensitivity effects is related to solar radiation; the terrestrial solar spectrum at unit air mass extends from about 295 mm to the infra-red region. The potent UV-B region of this spectrum extends from 295 to 315 mm, and the UV-A region from 315 to 400 nm. In spite of the very low fraction of less than 6% of ultraviolet radiation component in sunlight, light-induced damage is the dominant factor determining the lifetime of polymers routinely used outdoors. Polymers such as poly(vinyl chloride) are inherently photolabile and their use outdoors is only possible due to the availability of excellent light stabilizer systems.

A detailed knowledge of the wavelength sensitivity of polymer photodegradation processes is useful from a practical standpoint. It provides basic information needed to select physical filters, light absorbers, or pigments which might be used in polymer formulations to limit their photodegradation. With the development of improved techniques it might also be useful in estimating the lifetimes of polymers under different light sources such as extraterrestrial sunlight, indoor light sources, or under spectrally-altered light conditions. A pertinent example of the latter is the assessment of photodegradation of polymers under sunlight which is spectrally-altered due to partial depletion of the stratospheric ozone layer. Reduced ozone levels increase the UV-B levels in sunlight reaching the earth's surface. The ozone layer absorbs the UV-C and most UV-B in the extra-terrestrial sunlight passing through it to reach the earth's surface. Most biological processes as well as polymer damage is caused by UV-B radiation. Estimating the impact of this increase in UV-B quantitatively requires the use of reliable spectral sensitivity data as well as dose-response information.

This review will address the issue of definitions, summarize the experimental approaches in studying wavelength sensitivity, and discuss the published data.

2 Definitions

Wavelength-sensitivity or spectral sensitivity of a polymer is a general term for the efficacy of various regions or wavelengths (usually in the UV-visible region) of the electromagnetic spectrum in causing a specific photoreaction. Photodamage is sometimes used in place of photoreaction, implying the photoprocess of interest to result in an economic loss as in the case of deterioration of plastics used in outdoor applications.

Wavelength-sensitivity data can be expressed in several different ways depending on the technique employed to generate the data, and the intended use of the information.

2.1 Action Spectrum

In photobiology, the amount of radiation required to evoke a specific level of response (or an action), plotted as a function of the irradiation wavelength, is called an action spectrum [12]. However, the wavelength dependence of the response per available or incident unit of radiant energy has also been referred to as the action spectrum for biological phenomena. In most polymer studies, the dependence of the effectiveness, $\Im(\lambda)$, of incident radiation in causing a photoreaction or photo-damage, on the wavelength of irradiation, is considered to be the action spectrum. Effectiveness is defined here as the photoreaction per unit incident photon (or for constant photon fluence)

$$\text{effectiveness } \Im(\lambda) = \frac{\text{magnitude of reaction (or damage)}}{\text{number of incident photons}}.$$

It is conveniently expressed as a plot of photoreaction per fixed number of photons vs the wavelength of radiation to which the sample was exposed. The use of a fixed total photon fluence at each wavelength avoids the need to make the assumptions involved in calculating effectiveness on a per photon basis. Action spectrum will therefore not depend on the spectral irradiance distribution of the source used in its determination.

This definition is based on incident photons (radiation incident on the surface of a solid polymer or on the surface of a cell carrying a solution of the polymer). No corrections are made for scattering losses, back reflection, absorbance, or internal reflection effects. The action spectrum, therefore, depends on absorption characteristics of the polymer as well as on the quantum yield of the degradation process studied.

The action spectrum for surface yellowing of a mechanical pulp newsprint paper sample at ambient temperature in the wavelength region of $\lambda = 280\text{–}600$ nm is shown in Fig. 2. Mechanical pulps, being unbleached, have about the same composition as the pine wood from which it was derived. The

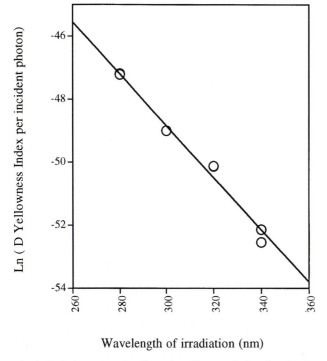

Wavelength of irradiation (nm)

Fig. 2. Action spectrum for change in yellowness index of newsprint paper (mechanical pulp from *Pinus taeda* species) [59]

lignin in pulp absorbs ultraviolet wavelengths and undergoes facile photo-yellowing. As the shorter wavelength radiation is relatively more energetic than the longer visible wavelengths, it is reasonable to expect greater damage per photon to be obtained at the ultraviolet end of the wavelength range. The typical dependence of photodamage on the wavelength of irradiation seen in Fig. 2 is consistent with this expectation, and the gradient of the plot conveniently quantifies the dependence. Several other polymer materials show this logarithmic dependence. The photon energy is a function of the reciprocal of the wavelength (for instance energy (kcal/mol) $= 2.865 \times 10^5/\lambda$ (Angstrom). However, in the wavelength region 280–600 nm employed in most action spectral studies, the dependence of photon energy on the wavelength can be well approximated by a logarithmic function. The line in Fig. 1 illustrates this logarithmic dependence. In most reported instances, the logarithm of damage per incident photon (or fixed number of photons) varies linearly with wavelength of irradiation, and therefore is a linear function of photon energy as well. Plots of the natural logarithm of average damage per constant photons vs the photon energy or the reciprocal of wavelength of irradiation (as opposed to the wavelength) are also linear. Recognition of the origin of the experimentally observed logarithmic dependence is important as it allows better interpretation of the empirical action spectral data in terms of the underlying photodegradation processes.

However, the wavelength dependence of the absorbance of radiation by the polymer (or the form of its absorption spectrum) may also be expected to contribute to the functional dependence of $\Im(\lambda)$ on wavelength. The relative importance of this factor can be assessed by comparing the action spectrum $\Im(\lambda)$ to that obtained using the number of incident photons weighed by the fractional absorption of the polymer at each irradiating wavelength. If the absorption characteristics of the polymer are mainly responsible for the observed logarithmic form of the action spectrum, this recalculation should yield a significantly different result. This was attempted with our data on yellowing of newsprint paper and poly(vinyl chloride) samples. In both cases, the gradient of the action spectrum was not significantly affected by the process. At least for these two systems the origin of the logarithmic form of the action spectrum cannot be attributed to the absorption characteristics of the polymer. The result, however, cannot necessarily be generalized to other polymer systems.

2.2 Activation Spectrum

It is a representation of the relative effectiveness of various regions of a polychromatic source spectrum in causing a given type of photoreaction or photodamage. The main difference between the action spectra and activation spectra is that only the latter is dependent upon the spectral irradiance distribution of the irradiating radiation.

Typically, it is a plot of the relative magnitude of photoreaction obtained in different spectral regions of the source spectrum, as a function of the irradiation waveband. Usually, the relative levels of damage attributed to each waveband is plotted, and absolute radiant energy is not used in construction of the spectrum. In common with all other measures of wavelength sensitivity, activation spectra are damage-specific, and, depending on the reaction studied, can even be duration-specific. Secondary reactions of accumulating photoproducts in the polymer can sometimes make wavelength-sensitivity data time-dependent. With solid polymers, photoproducts often concentrate on a thin surface layer a few microns thick [13], protecting the underlying layers from incident radiation [14]. This protective "filter effect" can affect the kinetics of the photoprocess and its effectiveness varies with the duration of exposure. With PVC for instance, the longer sequences of conjugated polyenes formed via photo-dehydrochlorination have high extinction coefficients [15] but are able to dissipate absorbed radiant energy as heat [14] without causing further photodamage in a sample that is heavily photodegraded. These factors, which complicate the wavelength sensitivity data, must be taken into account in extrapolating action or activation spectral data to other sources or exposure conditions.

Figure 3 shows the activation spectrum for yellowing of newsprint paper (unbleached mechanical pulp) exposed to a borosilicate-filtered xenon source.

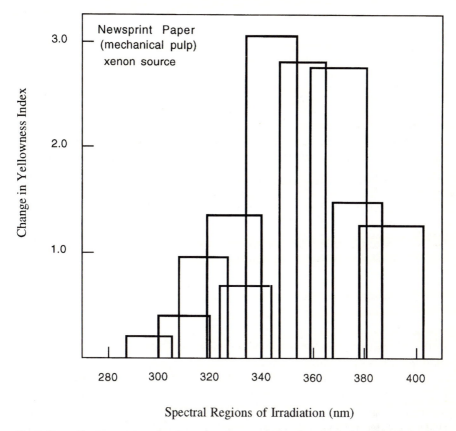

Fig. 3. The activation spectrum for change in yellowness index of mechanical pulp newsprint paper exposed to a borosilicate-filtered xenon source, for 13.8 h at 69 °C, determined by the cut-on filter technique [44]

The technique usually employed allows only the effect of fairly broad bands of the source spectrum to be studied. Hence the data is presented in the form of a bar diagram and in this case the wavelength interval of 334–355 nm shows the largest change in yellowness index. While the process studied (photoyellowing) and the polymer material used are the same as for the data in Fig. 2, the form of the spectrum is understandably very different. In Fig. 2 the data are expressed on a per incident photon basis and are not specific to any source, while in Fig. 3 the data are expressed on a per waveband basis and are specific to the filtered xenon source used in the experiment. The higher numbers of longer wavelength photons in the xenon source spectrum more than compensates for the effect of smaller numbers of higher energy ultraviolet photons reaching the sample during exposure. This allows higher relative photodamage to be obtained at wavelengths other than the shortest available from the source, provided of course the longer wavelengths are also absorbed by the polymer. The shape of the spectrum depends on the source spectrum, absorption spectrum of the sample, and quantum efficiency of the degradation process of interest.

Consider a source yielding a spectral irradiance $H(\lambda)$ at the sample position and an action spectrum of $E^*(\lambda)$ expressed in dimensionless units for a specific photoprocess occurring in the sample (with $E^*(\lambda) < 1$). The following expressions then indicate the effective irradiance and the dose rate available for that photoprocess. Spectral irradiance is defined as the irradiance (or radiant energy flux per unit area incident on a surface) per unit wavelength. Effective dose rate is sometimes also referred to as the instantaneous effective dosage, but the former term consistent with the usage in photobiology, will be used here:

$$\text{effective irradiance} = H(\lambda)E^*(\lambda) \qquad \text{(watts/m}^2\text{-nm)} \tag{1}$$

$$\text{effective dose rate} = \int_{\text{all }\lambda} E^*(\lambda)H(\lambda)\,d\lambda \quad \text{(watts/m}^2\text{).} \tag{2}$$

The total effective irradiance is essentially the instantaneous dose rate and its integration over the duration of exposure yields the effective dose:

$$\text{effective dose of radiation} = \int_{\text{all }\lambda} \int_0^T E^*(\lambda)H(\lambda)\,d\lambda\,dt \quad \text{(J/m}^2\text{).} \tag{3}$$

Two major assumptions are implicit in the above equations – the effect of individual wavelengths are assumed to be independent and additive, and the reciprocity rule is assumed to hold for the system. In instances where a photoprocess such as the discoloration brought about by short wavelengths is reversed by exposure to longer wavelengths, the additivity assumption clearly cannot apply. The reciprocity rule requires the photodamage due to a given dose of radiation to be independent of the radiant flux, variations in the flux, and the duration of exposure. Unfortunately, neither of these assumptions have been rigorously tested for polymer materials. Little or no data is available on the validity of the reciprocity rule in polymer photodegradation.

Since $E^*(\lambda)$ is only a relative weighting function applicable to a specific process, the effective dose rate and the dose calculated for different photoreactions will be numerically different. As such, these absolute numerical values for different photodegradation processes or for different types of polymers cannot be compared to each other in a meaningful manner. Experimental evaluations of wavelength sensitivity yield data expressed in units based on the measurement of damage and the units of irradiance. These might be converted to dimensionless form $\{E^*(\lambda)\}$ by expressing the photodamage as a fraction of the maximum damage obtained at any wavelength.

2.3 Wavelength Dependence of Quantum Yield

The quantum yield (or quantum efficiency) of a photoreaction is generally wavelength-dependent and differs from the above in that the reaction obtained per absorbed photon is used in its calculation. The quantum yield of main chain scission of a polymer, for instance, is the ratio of estimated scission events obtained per unit photon absorbed by the material.

$$\text{quantum yield } \Phi(\lambda) = \frac{\text{number of moles of photoproducts formed}}{\text{number of moles of photons absorbed}}.$$

The spectral distribution of quantum yields is also source-independent but photoprocess-specific. Figure 4 shows the wavelength-sensitivity of solvent-cast poly(methyl methacrylate) exposed to monochromatic radiation in terms of the quantum yield of scission [16]. These quantum yields were estimated from viscosity measurements assuming random main-chain scission. While quantum yields are particularly easy to determine for reactions in solution, they are somewhat difficult to obtain for opaque solids even when an integrated sphere is used to include scattered reflected fraction of light. This is particularly true of polymers where the chromophore is an impurity, present in low concentrations, and often having a low absorption coefficient. However, data obtained with solutions cannot be reliably extrapolated to the corresponding solid substrate. The observed quantum yields from irradiation of poly(vinyl chloride) solutions were very different from those of solid-state irradiations [13, 17], possibly due to different efficiencies of cage recombination of radical species participating in the photodegradation process.

In using quantum yield data to estimate photodamage, it is important to remember that $\Phi(\lambda)$ is not a fundamental material property. It may depend on factors such as film thickness, moisture content, and compounding/processing

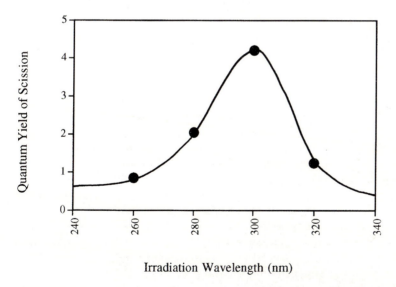

Fig. 4. Wavelength sensitivity of the estimated quantum efficiency of chain scission in poly(methyl methacrylate) exposed to near-monochromatic radiation. Based on data in [16]

conditions. Furthermore, quantum yields can be best interpreted only in the case of those photoprocesses involving well-defined, simple photoreactions.

The effective dose of radiation might be expressed in terms of the radiation *absorbed* by the polymer and the quantum yield of the particular photoprocess. For a polymer with a dimensionless spectral absorption $(1 - \exp(-k(\lambda))$, the action spectrum $E^*(\lambda) = (1 - \exp(-k(\lambda))\Phi(\lambda)$, and the effective dose can therefore be written as

$$\text{effective dose of radiation} = \int_{\text{all }\lambda} \int H(\lambda)(1 - \exp(-k(\lambda))\Phi(\lambda)\,d\lambda\,dt$$

$$(\text{J/m}^2). \qquad (4)$$

With photoprocesses where the absorption spectrum and the quantum yield spectrum are highly correlated, the following simpler expression applies:

$$\text{effective dose of radiation} = \int_{\text{all }\lambda} \int H(\lambda)\Phi(\lambda)\,d\lambda\,dt \quad (\text{J/m}^2). \qquad (5)$$

The expression in Eq. (5), generally used in photobiology, implies the absorbance spectrum of the substrate to be highly correlated to the quantum yield spectrum. While Martin et al. [18] have suggested that the expression at Eq. (4) is perhaps more appropriate for polymers containing photostabilizers and impurities, its practical superiority over the simpler form in Eq. (5) has not been demonstrated. It is feasible that in many polymer photodegradation processes the absorption factor merely scales the spectral quantum yield. The activation spectrum is then expressed as [18]

$$\text{activation spectrum} = H(\lambda)(1 - \exp(-k(\lambda))\Phi(\lambda) \qquad (6)$$

3 Experimental Approaches

Action and activation spectra can be experimentally determined from studies employing exposure of samples to either monochromatic or polychromatic UV-visible radiation. Rather lax use of these terms in the literature on wavelength sensitivity has led to some degree of confusion. It is convenient to define these terms in a manner consistent with most of the reported literature by making a clear distinction between the different experimental approaches.

An action spectrum shows the relationship between the effectiveness of a particular photoprocess and the wavelength of irradiation. They can be conveniently generated from the measurement of photodamage resulting from exposure of samples to near-monochromatic radiation of known irradiance. Action spectra can also be calculated from source-specific activation spectra, provided the irradiance distribution used in generating the data is accurately known. Accordingly, action spectra have been derived from polychromatic exposure data obtained using a series of cut-on filters [20]. (While no activation spectrum was reported in this particular study, their data were sufficient to construct one.) This conversion, however, involves several assumptions and practical limitations, discussed later. Data from exposure to near-monochromatic radiation uses narrow wavebands and are therefore less likely to be affected by complicating synergistic effects.

An activation spectrum shows the effectiveness of a particular photoprocess caused by different regions (or even individual wavelengths) of the polychromatic radiation used for irradiation. It is conveniently generated from polychromatic exposure experiments using a series of cut-on filters to isolate different wavebands of the irradiating spectrum. Activation spectra are likely to be more representative of the photodegradation obtained under exposure to polychromatic radiation, as the samples are exposed to the spectral band of interest and all longer wavelengths in the source spectrum. Provided the spectral irradiance distribution of the radiation at sample position is known, activation spectra can be calculated from action spectral data. The conversion involves the same set of limitations referred to above, and discussed later.

In both these definitions the term "effectiveness" is used to denote change in the extent of photoreactions or photodamage per constant incident photons per constant area of sample. The term "efficiency" is used exclusively in reference to absorbed radiation, as in the case of quantum efficiency. Table 1 summarizes these definitions, and illustrates the four main approaches to generating data.

Table 1. Summary table of definitions relating to wavelength sensitivity of polymers

Wavelength sensitivity

Wavelength sensitivity refers to the efficacy of various regions or wavelengths of the electromagnetic spectrum in causing a given photoreaction or a photoprocess. Data can be expressed either in relative or absolute terms.

1 Action spectrum: a plot showing the dependence of the effectiveness of a photoprocess on the wavelength of irradiation.

1.1 From exposure to monochromatic radiation: determined from experiments involving exposure of samples to near-monochromatic radiation.

(Sect 3.1)

1.2 From exposure to polychromatic radiation: calculated from data from exposure experiments using a series of cut-on filters and a specific polychromatic source.

(Sect. 3.3)

2 Activation spectrum: data showing the dependence of the relative extent of photoreaction upon the different regions of a source spectrum irradiating the sample.

2.1 From exposure to polychromatic radiation: determined from exposure experiments using a series of cut-on filters and a specific polychromatic source.

(Sect. 3.2)

2.2 a) From exposure to monochromatic radiation: determined from exposure of samples to polychromatic radiation that is spectrally dispersed.
b) Calculated from monochromatic action spectral data using the spectral irradiance distribution of a polychromatic source.

(Sect. 3.4)

3.1 Determination of Action Spectra Using
   Monochromatic Exposure

3.1.1 Interference Filters

The simplest experimental technique is to expose a sample of the polymer to near-monochromatic radiation and to measure the change in the property of interest before and after exposure. Pure monochromatic radiation is not obtainable in practice, and at best only a narrow waveband with a finite bandwidth can be isolated. Such a waveband can be isolated from a source spectrum using an interference bandpass filter [21, 22]. These are essentially Fabry-Perot interferometers [23] of fixed spacing. Wavelength intervals of a fraction of a nanometer can be obtained with the best filters but only at the expense of the fraction of light transmitted. A good compromise is a filter with a half-band width of about 15 nm. Due to thermal expansion effects on the optics of the cavity, their spectrum is usually temperature-dependent ($\sim 10^{-2}$ nm/°C for filters from Melles Griot, Arnhem, Holland). A spectral wedge or a continuous interference filter which transmits different wavelengths at different regions along its length might also be used. Either a filtered xenon source or a mercury vapor source is popularly used with an appropriate interference filter set.

   Alternatively, it is possible to use a laser beam such as the 488 nm argon ion laser beam, appropriately diverged as a monochromatic source [24]. A dye laser which can yield different wavelengths is convenient, but the monochromatic radiation output is often very low, particularly at UV wavelengths. At high intensities of laser radiation, ablative degradation can occur charring the polymer [25].

3.1.2 Spectrograph

A dispersed spectrum of a suitable polychromatic source can also be used as a near-monochromatic source to obtain action spectra. A quartz prism or a grating spectrograph is used to disperse the source spectrum and the samples are placed in the appropriate region of the spectrum. A monochromator [26, 27] used with a suitable source allows a part of this dispersed spectrum to be isolated at a narrow exit slit to yield a good monochromatic source.

   Perhaps the best example of a sophisticated spectrograph is the Okazaki Large Spectrograph (The National Institute for Basic Biology in Okazaki, Japan) [28]. It uses a powerful xenon source (30 kW) and a double-blazed grating (1200 lines/mm) to project a dispersed spectrum onto a 10-m-long focal curve. At various sample positions along this curve, near-monochromatic radiation of wavelength between 250 and 1000 nm can be obtained at a resolution of less than 1 nm/cm. With adequate power controls, continuous monitoring of source performance and recording of drifts in intensity at the sample position, the equipment is able to deliver well-defined near-monochromatic radiation over fairly long periods of time. A simplified diagram of the system is shown in Fig. 5. It is possible to change the path of radiation by reflecting it off a UV-reflecting plane mirror to allow flat samples placed horizontally to be exposed.

Two basic exposure modes can be employed to obtain action spectra using a near-monochromatic source: exposure of a set of samples to constant photon fluence (equal numbers of photons at each wavelength) or exposure of all samples for the same duration at each wavelength. In the latter case, the photon fluence obtained at different wavelengths will be very different and will depend upon the spectral irradiance distribution of the source. To calculate the effectiveness $\mathfrak{I}(\lambda)$ in units of damage per incident photon, the ratio of the measured change in property of interest ΔP to the number of incident photons is calculated for each irradiation wavelength. This procedure assumes linearity in the photochemical response of the sample over the range of photon fluence rates encountered in the exposure experiment. The alternative method of exposing to constant photon fluence at each wavelength is more attractive but requires different exposure durations at different wavelengths to ensure constant total photon fluence. In the Okazaki spectrograph system, the durations of exposure are monitored and conveniently controlled by a computer system.

Samples after exposure are stored in the dark until the appropriate measurements are carried out. Several photoreactions of interest such as yellowing discoloration of poly(vinyl chloride) are affected by "dark reactions" [29, 30]. Others, such as the light-induced discoloration of a poly(isophthalamide) are reported to be unaffected by dark reactions [31]. Careful attention must therefore be paid to the duration of storage as well as to the storage conditions of the exposed samples.

3.1.3 Action Spectra Using a UV-Enhanced Source

It is possible in theory to generate an action spectrum which takes into account the effect of incremental monochromatic radiation in excess of the spectral irradiance of a source spectrum. Narrow-band, near-monochromatic radiation might be superimposed over the polychromatic spectrum from a xenon source to obtain radiation spectrally enhanced in any selected waveband. Using a series of interference filters, a series of cut-on filters, and two filtered xenon sources, it should be possible to generate an action spectrum for polymer degradation under spectrally-altered, UV-enhanced (or even visible-enhanced) source spectra. The effective radiation and dose rate in such an experiment can be written as follows:

$$\text{effective irradiance} = H(\lambda)\aleph(\lambda)E^*(\lambda) \quad \text{(watts/m}^2\text{-nm)} \tag{7}$$

$$\text{dose rate} = \int_{\text{all}\lambda} H(\lambda)\aleph(\lambda)E^*(\lambda)\,d\lambda \quad \text{(watts/m}^2\text{)} \tag{8}$$

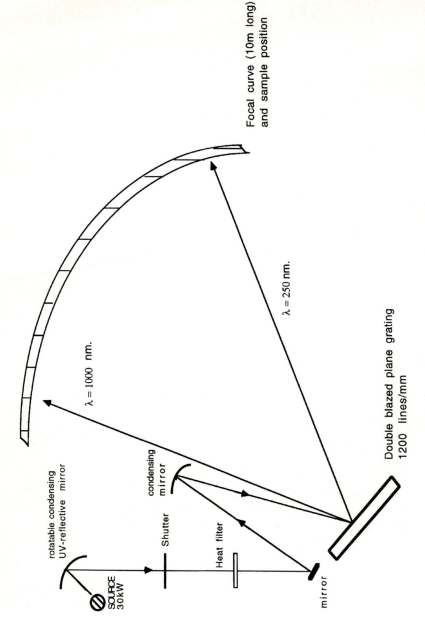

Fig. 5. Basic features of the large spectrograph system (NIBB, Okazaki, Japan). A second 6 KW source used for low-power exposures and the mercury-vapor sources used for calibration are not shown. Diagram based on [28]

where $\aleph(\lambda)$ is the weighing function which takes into account the narrow band of radiation from secondary source superimposed on the base xenon spectrum.

While this type of experiment has never been reported, it can provide unique and useful information. For instance, it can be useful in determining the effects of slightly increased levels of UV radiation in sunlight on its potential for increased polymer photodegradation. Levels of UV-B in terrestrial sunlight are expected to increase as a result of the partial depletion of the stratospheric ozone layer. The effect of this spectrally altered, UV-B-rich sunlight on polymer degradation can also be ascertained from activation spectra, subject to several assumptions mentioned earlier.

3.2 Determination of Activation Spectra From Polychromatic Exposure Experiments

As activation spectra are source-specific and can be obtained by more than one experimental approach, it is useful to specify the following consistently when discussing a spectrum: i) the specific source used, ii) the photoprocess studied, preferably with the measurement used, and iii) the technique (cut-on filter, light dispersed by a grating, light dispersed with a prism) used in the determination.

3.2.1 Cut-on Filter Technique

This technique relies on a series of sharp glass cut-on filters to determine the effect of exposure of polymer samples to narrow wavebands of a source spectrum [31–33]. Early work used sheets of various thermoplastics as cut-on filter materials [34]. Figure 6 shows transmittance spectra, $T(\lambda)$, of a series of glass cut-on filters typically used for this purpose. Figure 6 also shows $(T_i(\lambda) - T_{i+1}(\lambda))$ for a selected pair of cut-on filters ($i = 5$ and 6), the incremental spectral band transmitted, and the corresponding increment of radiation transmitted from a filtered xenon source. The filtered xenon source is popularly used in photodegradation experiments because it is designed to emit a spectrum similar in irradiance distribution to that of terrestrial sunlight.

$$\text{Increment of radiation } (\lambda) = H(\lambda)(T_i(\lambda) - (T_{i+1}(\lambda)) \tag{9}$$

where $H(\lambda)$ is the xenon source spectral irradiance distribution. The increment is the spectral band of radiation transmitted by filter 5 but blocked by filter 6.

A sample of the polymer is placed behind each filter in the series and the whole set exposed to a source spectrum simultaneously. Each sample receiving radiation filtered by a different filter in the series will undergo photodamage at a slightly different rate (and sometimes even via a different mechanism). Following a suitable duration of exposure the samples are removed and the light-induced damage obtained is determined using an appropriate measurement.

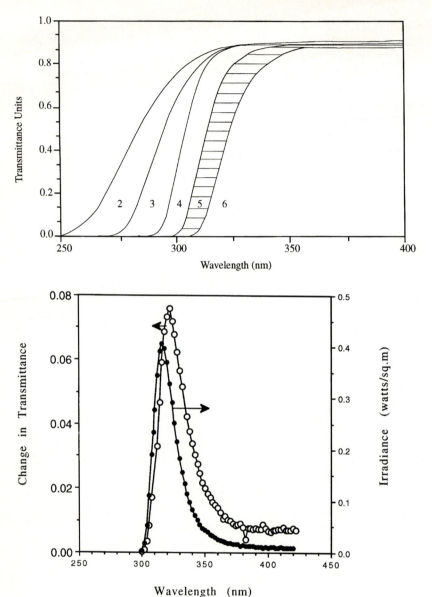

Fig. 6. Transmittance characteristics of cut-on filters typically used in determination of activation spectra. The waveband transmitted by a pair of filters was calculated by spectral subtraction of transmittance curves; the corresponding irradiance increment was calculated using the spectral irradiance data for a filtered-xenon source

The difference in degradation between samples exposed behind each pair of filters is ascribed to the radiation in the band of wavelengths that is represented by the incremental spectral band of the filter pair (see Fig. 6). The spectral regions are indicated in the activation spectrum by the range of wavelengths at a 20% transmittance level of the incremental transmittance band, which is approximately the width at half the peak height. The incremental damage of the sample exposed behind the shorter wavelength cut-on filter of the pair is obtained, normalized, and plotted in the form of a bar diagram as a function of the wavelength range transmitted by the respective filter pair. The activation spectrum is therefore a bar diagram and shows the regions of the source spectrum which cause the maximum damage. Its resolution depends on the number of filters used to generate it. Figure 4 shows an activation spectrum for change in yellowness index of newsprint paper exposed to filtered xenon radiation, constructed in this manner.

The activation spectrum is the plot of change in a property, (Δ Damage), before and after exposure of identical samples behind filters i and i + 1 as a function of the bandpass $(T_i(\lambda) - T_{i+1}(\lambda))$. Conventionally, the bandpass is indicated by wavelengths corresponding to 20% of the maximum in the increment of transmittance of the relevant filter pair. However, the use of bandpass based on the increment in radiation (as in Eq. (9)) of filter pairs in the construction of activation spectra is more meaningful. The bandpass calculated using the increment in radiant energy transmitted by a filter pair tends to be wider than that obtained using the increment in transmittance.

In some instances, where a high variance in wavelength-dependent photochemical changes is obtained, it is more convenient to present the data as the change in absolute value of a selected property, as a function of the cut-on wavelength of the filter. Figure 7 shows a spectral sensitivity curve for wavelength dependence of the tensile elongation at break of an ethylene copolymer film exposed to a borosilicate filtered xenon source. The curve shows the cumulative effect of wavebands of the source spectrum on the measured property. Conventionally, data presented in this manner is not referred to as an activation spectrum. It is referred to here as the cumulative spectral sensitivity curve.

3.3 Action Spectra Derived From Data From Polychromatic Exposure Experiments

The interconversion of action and activation spectral data is possible, but invariably involves several assumptions that may affect its validity. These include:

i) assumption of the law of reciprocity (sometimes called the Bunsen-Roscoe Law);
ii) assumption of additivity of effects of different wavelengths (or the lack of synergistic or antagonistic effects); and
iii) assumption of time-independence of absorbance characteristics of the polymer.

With most polymers the information needed to assess the validity of these assumptions is not available at this time. Interconversion of data is not

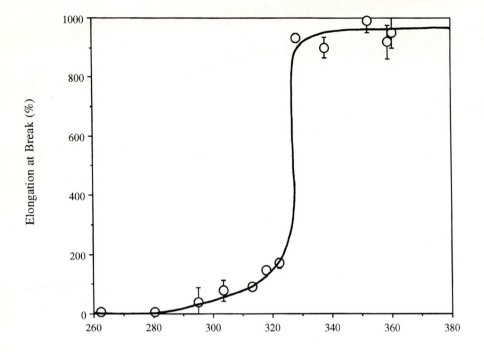

Cut-on Wavelength of Filter (nm)

Fig. 7. A cumulative spectral sensitivity curve for the decrease in elongation at break of (ethylene-carbon monoxide 1%) copolymer sheets on exposure to a filtered xenon source for 219 h at 60 °C, determined by the cut-on filter technique. The error bars show the standard error of the mean [133]

a straight-forward exercise and may not always even be applicable. The validity of a converted or derived spectrum can only be established by comparing it to one determined experimentally. In reporting wavelength sensitivity data it is therefore crucial to indicate clearly the experimental approach used, as well as data obtained by calculation, as the data from different techniques need not always agree.

3.3.1 Action Spectra Calculated from Activation Spectra

The extent of photodamage obtained on exposure of a polymer depends upon the available radiant energy at the sample surface and the action spectrum $E(\lambda)$ for the process. Note that the action spectrum $E(\lambda)$ is no longer dimensionless as $E^*(\lambda)$ in Eqs. (1) and (2) but have units depending on the damage studied. The integral of Eq. (9) can be used along with values of incremental change in (Δ Damage) between adjacent filters to calculate the effectiveness E_{i-j}, and therefore the action spectrum for the process.

$$E_{i-j}(\lambda) = \frac{\Delta(\Delta \text{ Damage})}{\iint I_0(\lambda)(T_i(\lambda) - T_j(\lambda)) \, d\lambda \, dt}$$ (10)

where (Δ Damage) is the change in property as a result of exposure to radiation, and I_0 is the intensity of incident radiation with dimensions of (energy) (time^{-1}). The calculated action spectrum is the plot of relative values of $E_{i-j}(\lambda)$ as a function of the bandwidth of the relevant wavelength interval. These values of $E_{i-j}(\lambda)$ are approximate functions of the maximum in bandpass obtained with each filter pair. However, the activation spectrum not being constant within the bandpass defined by the pair of filters, best fit for the data need not be necessarily obtained at the maximum value of $(T_i(\lambda) - T_j(\lambda))$. An iteration procedure might be used to obtain the wavelength corresponding to best fit of datapoints [20]. This procedure of denoting the narrow wavebands in the activation spectrum by an average single wavelength is a limitation of this approach. The only additional information needed to derive an action spectrum from an activation spectrum is the spectral irradiance distribution for the irradiation. This is measured in terms of intensity at the sample position using a radiometer.

The action spectrum for yellowing of poly(vinyl chloride) formulations, based on monochromatic exposure data [35, 36], as well as an activation spectrum for yellowing generated using the filter technique [37], have been reported by Andrady et al. An action spectrum was derived from this activation spectrum using the above approach. In Fig. 8 the experimentally determined action spectra for yellowing of compounded, extruded, rigid PVC formulations is compared to one calculated from the activation spectrum. The area under the increment in radiation curve for each pair of filters was converted to an equivalent number of photons using the peak wavelength of the increment. The action spectrum (change in yellowness index per incident photon) was calculated using this estimate of the number of photons, and the incremental photodamage data. Both the experimental and calculated spectra show a high degree of linearity ($r^2 > 0.93$), but the gradient of the spectrum calculated from the activation spectral data is about 30% larger than that of the action spectrum from direct determination.

3.3.2 Wavelength Sensitivity of Quantum Yields Calculated from Cut-on Filter Technique

Activation spectra might be converted into quantum yield spectra by taking into account the light absorbed by the polymer sample, as given by the Beer-Lambert Law. In this case, the amount of light absorbed throughout the bulk of sample (as opposed to that incident on the sample) is calculated.

Provided the absorbance of the sample is known, the light intensity at a depth x in a sample of thickness d can be readily calculated using Beer's Law.

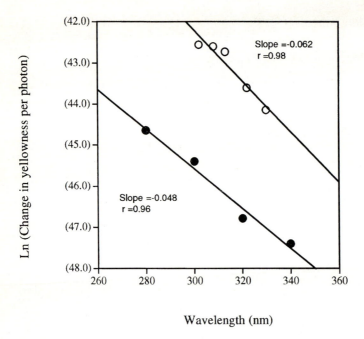

Fig. 8. A comparison of the experimentally determined action spectrum (lower plot) with one calculated from an activation spectrum (*upper plot*) for yellowing of a poly(vinyl chloride), PVC, samples containing 2.5 phr of rutile titania

As with Eq. (2) above, one can write the effective irradiance at a depth x as follows:

$$\text{effective irradiance} = \int_{\text{all }\lambda} H(\lambda, x)S^*(\lambda)\,d\lambda$$

$$= \int_{\text{all }\lambda} I_0(\lambda)\,10^{-(\Sigma\varepsilon_i(\lambda)c_i)x}\,S^*(\lambda)\,d\lambda \qquad (11)$$

where $H(\lambda, x)$ is the radiant energy at wavelength λ, and at a depth x, $\varepsilon_i(\lambda)$ the molar extinction coefficient, c_i the concentration of absorbing species, and $S^*(\lambda)$ quantum yield spectrum expressed in dimensionless form ($S^*(\lambda) = S(\lambda)/S(\lambda)_{\text{max}}$). Since the Beer-Lambert term in Eq. (11) takes into account the radiation absorbed by the polymer, the spectrum $S^*(\lambda)$ is not obviously the same as $E^*(\lambda)$ in Eq. (2) based on incident radiation. While quantum yield can strictly be applied only to well-defined photoreactions of known mechanism, it is used here in a more general sense to denote the efficiency of absorbed quanta in yielding a given photodamage or a reaction product. The effective dose is then written as follows:

$$\text{effective dose} = t \int_{\text{all }\lambda} I_0(\lambda)(1 - 10^{-(\Sigma\varepsilon_i(\lambda)c_i)d})/(\Sigma\varepsilon_i(\lambda)c_i)\,S^*(\lambda)\,d\lambda \qquad (12)$$

where t is the duration of exposure.

With the same approach used to obtain Eq. (10) but using Eq. (12) to describe the dose of radiation, an alternative expression analogous to Eq. (10) can be obtained as follows. The incremental damage (Δ Damage) is again assumed to be a linear function of the effective dose of radiation to obtain Eq. (13):

$$S_{i-j}^{*} = \frac{\Delta(\Delta \text{ Damage})}{t \int\limits_{\text{all } \lambda} I_0(\lambda)(T_i(\lambda) - T_j(\lambda))(1 - 10^{-(\Sigma \varepsilon_i(\lambda)c_i)d})/(\Sigma \varepsilon_i(\lambda)c_i) \, d\lambda}. \tag{13}$$

A quantum yield spectrum of the form described by Eq. (13) was reported for polystyrene [20]. This base spectrum and the effective dose expression in Eq. (12) were used to estimate the expected damage for polymers of known absorbance. Working with transparent polymer formulations containing 13 different light-stabilizer systems of known absorption coefficients (and two light sources), Allan et al. [20] showed the experimentally-obtained rates correlated well (r = 0.95) with the expected values.

This procedure is applicable to photoprocesses which vary linearly with incident energy and where the light-absorber (the stabilizer which acts as a filter) is not itself degraded and is uniformly distributed in the polymer. While the model systematically underestimated the magnitude of degradation, it successfully predicted the rates of degradation over all conditions studied. The invariance of absorber concentration in time implicit in these equations is unrealistic as some depletion of the light absorber, at least in the surface layers, must invariably occur.

The treatment is general enough to be applied with attenuation functions other than that based on Beer's Law. For instance, a polymer containing a finely dispersed light-absorbing opacifier such as titania might be considered. Attenuation of radiation in such a system was discussed by Andrady and Shultz [38, 39] and is due to shielding of the underlying polymer due to absorption of light by the particulate filler, as well as due to absorption by the polymer matrix itself.

Consider a polymer with an absorption decrement of $k_p(\text{cm}^{-1})$ containing a UV absorber with an absorption decrement of $k_a(\text{cm}^{-1})$ uniformly dispersed within it. It also contains a volume fraction V of a polymer-insoluble radiation-absorbing pigment composed of monodisperse spheres of radius r. Effective irradiance at the depth x for the system can be written, based on Andrady and Shultz [38, 39], as follows:

$$\text{effective irradiance} = \int\limits_{\text{all } \lambda} I_0(\lambda)(1 - 3V/2)^{x/2r_3} \exp(-(k_p + k_a)x)S^*(\lambda) \, d\lambda$$

$$\tag{14}$$

where r_3 is the area-average radius of the opaque particles, and I_0 is the intensity of radiation incident on the sample surface. As the attenuating effect of pigment particles might be expected to overshadow that due to absorbance by the matrix, the above equation might be simplified by disregarding the exponential term:

$$\text{effective irradiance} = \int\limits_{\text{all }\lambda} I_0(\lambda)(1 - 3V/2)^{x/2r_3} S^*(\lambda) d\lambda. \tag{15}$$

Integrating the expression for effective radiation at any wavelength over the thickness of the sample L, as with Eq. (12), the total effective dose of radiation can be obtained as follows:

$$\text{effective dose} = t \int\limits_{\text{all }\lambda} I_0(\lambda) 2r_3 \{\mathfrak{R}^{L/2r_3} - 1\} \{\text{Ln } \mathfrak{R}\}^{-1} S^* d\lambda \tag{16}$$

where $\mathfrak{R} = (1 - 3V/2)$.

The difference in the extent of net photoreaction in samples exposed behind adjacent filters, i and j, $\Delta(\Delta \text{ Damage})$, and Eq. (16) above, can then be used to estimate action spectra using an equation analogous to Eq. (13). Again, a linear relationship between the effective dose and (Δ Damage) is assumed:

$$S^*_{i-j} = \frac{\Delta(\Delta \text{ Damage})}{t \int\limits_{\text{all }\lambda} I_0(\lambda)(T_i(\lambda) - T_j(\lambda))\{4(r_3)^2(1 - 3V/2)^{L/2r_3} + \mathfrak{R}\} L\mathfrak{R}^{-2} d\lambda}. \tag{17}$$

Neither the Beer Lambert law nor the simple expression for light shielding by opaque particles take into account backscattering or specular reflection of light by the sample. More exhaustive treatments based on phenomenological theories (such as the Kubelka-Munk equation [40]) can be employed to derive even more sophisticated expressions. Complex expressions for transmittance of light through biological media have been derived using such approaches [41].

3.4 Activation Spectra from Monochromatic Exposure Experiments

3.4.1 Spectrographic Technique

A collimated beam of light spectrally dispersed by a diffraction grating or a quartz prism yields a spectrum which can be used as a series of wavebands of near-monochromatic radiation. The intensities of different regions of the dispersed spectrum depend on the spectral distribution of the source. The dispersed spectrum is focused only on a few centimeters of the sample surface, and exposure is carried out for a constant duration. A microspectroscopic technique is generally used to study changes in absorbance properties of the sample subsequent to exposure.

A line spectrum of a Hg vapor source (or other line source) refracted through the same dispersing prism and recorded on a photographic plate placed at the sample position is used to locate accurately the different wavelengths across the surface of the sample. As the refractive index of quartz is not a linear function of the wavelength, the angular dispersion or the resolution of the spectrum will not be linear over the sample surface. This requires meticulous care in placing the sample and in establishing the location of several discrete wavelengths on its surface. Substituting a diffraction grating in place of the prism avoids some of these difficulties.

Using the data obtained in this manner to generate an activation spectrum involves the assumption that any losses of radiation during dispersion and focusing, affect all wavelengths uniformly. However, the activation spectra generated from monochromatic exposure can be expected to agree well with those obtained from polychromatic exposure experiments only in instances where the effect of different wavelengths are additive. With most common polymers the approach yields activation spectra in good agreement with those from cut-on filter technique [19]. This indicates not only the additivity of effects but also the lack of synergistic or antagonistic effects at different wavelengths, for these photoprocesses.

Hirt and Searle [19] and Johnson et al. [31] reported the use of a borosilicate-filtered xenon source dispersed with a quartz prism spectrograph to obtain activation spectra of several polymers. The 290–410 nm region of the spectrum was dispersed and focused onto about 1 cm length of a strip of polymer. Solar radiation, followed with a heliostat and focused onto the entrance slit of the spectrograph, was also used as an exposure source for some of the data. Following exposure to radiation for a constant duration the sample was scanned in appropriate spectrophotometric equipment to measure absorption of radiation (ultroviolet, visible or infra-red), changes in fluorescence, or changes in yellowing, at different points across the exposed area of sample (i.e. corresponding to different wavelengths of exposure). The exposed sample was moved by a motorized drive across a 0.5-mm slit placed in the exit beam of the spectrophotometer. They successfully identified regions of the xenon source spectrum causing maximum changes in selected properties of the polymer. Data by Hirt and Searle [19] summarized in Table 2 below have been sometimes quoted without clearly stating that the data pertains solely to terrestrial solar or solar-simulated radiation and neglecting to emphasize their specificity to the particular photoprocesses they address, namely changes in the absorption of UV/visible/IR radiation, or fluorescence emission. Activation spectra cannot be generalized to cover different sources or photodegradation processes other than the one used in its generation.

As indicated in Sect. 3.3, it is possible in theory to calculate the activation spectral response of a photoprocess from the known action spectrum for that process and the detailed spectral irradiance distribution of the source of interest. The various assumptions involved in such a calculation have already been pointed out.

Table 2. Regions of maximum effectiveness in causing photoreactions in polymers exposed to simulated solar radiation, determined using laboratory spectrograph techniques [19]

Polymer	Monitoring technique[a]				Region of maximum
	UV	Vis.	IR	Fluo.	effect (nm)
1 Polyesters		*			325
2 Polystyrene		*		*	319
3 Polyethylene			*		300
4 Polypropylene			*		370
5 Poly(vinyl chloride)		*		*	320
6 Poly(vinyl acetate)	*	*			< 280
7 Polycarbonate	*	*		*	285–305
8 Cellulose acetate butyrate	*	*			295–298
9 Styrene acrylonitrile copolymer	*	*			290–325

[a] Spectroscopic techniques used to monitor the extent of degradation is shown in this column. UV – ultraviolet absorption, Vis. – visible light absorption, IR – infra-red absorption, and Fluo. – fluorescence spectroscopy

3.5 Experimental Factors that Can Influence Wavelength Sensitivity Data

With activation spectra directly determined using either approach, it is important to select an appropriate duration of exposure for the experiment. The duration should be long enough to obtain a measurable difference in the property of interest for samples exposed under adjacent filters. Yet it must not be long enough to induce any photoprocesses not normally obtained in exposure of samples to common sources of radiation. For instance, with light-induced yellowing discoloration of poly(vinyl chloride), the photoproducts formed at very high extents of reaction may act as a protective surface filter retarding photodegradation [16, 17]. Where the substrate is a plastic product such as a packaging film, the duration might be selected to match the highest levels of photodegradation obtained during its useful lifetime. As seen from the curves in Fig. 9 showing the kinetics of yellowing of rigid titania-free poly(vinyl chloride) formulations extruded into sheets [42], the relative damage under individual filters can indeed be a function of the duration of exposure. In this case, the relative effectiveness of different spectral bands changed with the duration of exposure, but the region of the spectrum associated with maximum yellowing was about the same at all durations.

A second important factor is the thickness of samples. As shown by several workers [32, 43], thicker samples absorb more of the incident radiation and will undergo degradation at a faster rate. In the case of an aromatic type polymer, the thicker samples are also able to absorb longer wavelengths of radiation. In the case of polyarylate photodegradation the dose needed to change the transmittance at 420 nm by 20% was studied as a function of the wavelength of irradiation [43]. The action spectrum shifted by 20 nm to longer wavelength when the sample thickness was varied from 70 to 1830 microns. Earlier data

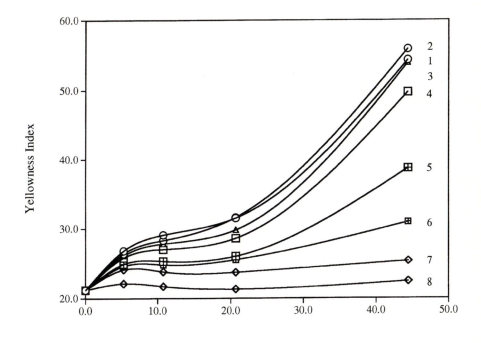

Fig. 9. Kinetics of yellowing of unstabilized rigid PVC formulations exposed behind different cut-on filters to filtered xenon source radiation [42]

[32] by Searle for polyester films 75 and 2000 microns thick also illustrate the same. Some modes of damage, such as surface discoloration of opaque formulations, are independent of the sample thickness while others such as changes in impact strength are likely to depend on sample thickness.

Depending on the energy of activation associated with the process, photodegradation is accelerated at higher temperatures. This may also cause a shift in the action spectrum of a polymer, as shown by the action spectrum for change in transmittance of a 700 micron polycarbonate film [43]. A change in temperature from 10 to 50 °C shifted this spectrum to longer wavelength by 10 nm. These shifts are of particular importance in establishing wavelength thresholds for photodegradation.

The temperature often determines the shape of action or activation spectra where both thermal and photochemical processes produce the same outcome. Both lignocellulose and poly(vinyl chloride), for instance, yellow on exposure to UV radiation as well as on heating. In the case of the latter substrate, even the mechanism of yellowing is the same. It is, therefore, necessary to include adequate controls in the experiments to estimate the relative importance of thermal processes in those exposures that cause combined photo-thermal degradation. Alternatively, additional samples might be exposed behind filters with long-wavelength cut-on values allowing no UV or visible radiation to reach the sample. In the case of newsprint paper exposed to a filtered-xenon source for 13.8 h at 69 °C, the thermal contribution to yellowing was estimated at about 17% [44]. These estimates were made by comparing the yellowing obtained with samples exposed behind no filters, cut-on filters, and special filters which filtered out most of the UV-visible radiation but transmitted the infra-red component of the source radiation.

4 Wavelength Sensitivity of Naturally-Occurring Polymers

4.1 Ligno-Cellulose

Lignin, which comprises 29–33% by weight of softwood [45], contains numerous chromophores [46] and is a very efficient absorber of ultraviolet radiation and visible light. As much as 80–95% of the absorption coefficient of wood can be ascribed to the lignin fraction [47]. The complex photochemistry of yellowing of lignin-containing materials on exposure to UV-visible radiation is not completely understood; the present understanding of the process was succinctly summarized recently [48]. Phenoxy radical precursors can yield yellow quinone compounds via at least four pathways and a detailed discussion of the various mechanisms has been presented [49–55].

Studies on the wavelength sensitivity of lignocellulose dates back to the 1940s and include studies by Nolan et al. [56] and Van der Akker et al. [57]. The latter reported a relative spectral sensitivity curve for spruce groundwood in the wavelength region of 200–385 nm. Since then, the spectral sensitivity of paper, pulp and wood to yellowing and/or brightness reversion on exposure to UV and visible radiation has been reported by many workers. The source of lignocellulose introduces some degree of variability into the data. Most recently reported action spectra are for yellowing of refiner pulp [58] and groundwood pulp [59]. In the latter study employing near-monochromatic radiation, several wavelengths in the interval 280–600 nm were used, both with exposure for a constant duration and to a constant number of photons. Action spectra (see Fig. 2) from the two techniques showed similar gradients, suggesting linearity of yellowing to be a function of intensity in the range of intensities pertinent to the exposure. Exposure to wavelengths of 500 and 600 nm did not result in yellowing of the paper; instead a slight bleaching effect was observed.

The activation spectrum for yellowing and decrease in brightness of newsprint paper exposed to xenon radiation has been obtained using the cut-on filter technique [44]. As seen in Fig. 4, the wavelength interval 335 to 380 nm was mostly responsible for both yellowing and brightness reversion of paper.

4.2 Wool

Exposure of wool keratins to sunlight is well known to cause yellowing [60, 61], bleaching [62], and main chain scission of the proteins [63]. Launer [61] established that visible radiation in sunlight causes photo-bleaching of wool, and that the UV region causes yellowing. Depending on the relative levels of these two components in sunlight at a given location or in a given season, net yellowing or net bleaching of wool may result.

As with lignins, the photochemistry of yellowing has not been fully elucidated, but the available data strongly suggests reactions of tryptophan residues to be mainly responsible for the process. Oxidation of the aromatic ring of the amino acid to hydroxytryptophan and the subsequent rupture of the indole ring to form 3-hydroxykynurenine, a yellow pigment [64], is a possible pathway. Kynurenine has been isolated from yellowed wool [65], and the yellowing of 29 types of keratins correlated well with their tryptophan content [66]:

3-hydroxykynurenine (yellow)

On extended irradiation, however, all wools will invariably yellow at about the same rate [67]. Lennox et al. reported an action spectrum for yellowing of wool with and without a fluorescent brightener (1% of a bistriazinyldiaminostilbene) [68]. From their reported data, an action spectrum of the following form can be derived:

$$\text{Ln}(\Delta \text{ Yellowness Index}) = -0.025 \, \lambda + 9.9 \quad r^2 = 0.95. \tag{18}$$

However, their paper does not describe in detail the manner in which the wavelength sensitivity data was generated. When the data was apparently corrected for the spectral irradiance in sunlight (to derive an activation spectrum), the maximum yellowing was estimated to occur at the interval of $\lambda = 340–420$ nm. The long wavelength region of $\lambda = 500–520$ nm yielded a secondary yellowing maximum.

5 Wavelength Sensitivity of Synthetic Polymers

As the photodegradation of a polymer depends on its formulation and is also influenced by processing, a bewilderingly large number of possible substrates exists. However, most reported studies have focused on a fairly small number of these, mostly the commodity thermoplastics and their compounds. Most have addressed the case of pure polymer; only a few have included typical formulations or compounds used in fabrication of common products. The photodegradation processes studied have included those of practical relevance such as discoloration, and those of fundamental significance such as changes in the average molecular weight of polymer. In the following selection of reported data, emphasis was placed on common polymers exposed to either sunlight or to filtered xenon sources designed to simulate sunlight.

5.1 Poly(viny chloride), PVC

PVC is well known to undergo yellowing readily on exposure to sunlight [69], and rutile titania (opacifier) is compounded into rigid PVC formulations used in extruded profile manufacture at a level of 7–13 parts per hundred to control premature photoyellowing [70–72] on routine outdoor exposure. These profiles are used as siding or door/window frames in the building industry. Early studies on the polymer showed the light-induced dehydrochlorination (which ultimately leads to yellowing) to be much slower in the solar UV region compared to that at shorter wavelengths such as 254 nm [16, 73, 74]. This reaction produces sequences of conjugated polyene structures which have high extinction coefficients for UV wavelengths [29, 75]. The longer sequences of polyenes (n > 8) absorb in the visible blue region of the spectrum causing the material to appear yellow [76]. The reaction is localized in the surface layers of the polymer [14], especially in opaque formulations. Rate of light-induced dehydrochlorination is reported to be first order with respect to the intensity of light [77, 78] and has a temperature coefficient [79] of 8–18 kJ mol^{-1}. The rate of degradation can, therefore, be affected by temperature, particularly by heat build-up [80] in pigmented formulations, or due to absorption of infra-red radiation by highly discolored samples [81] exposed outdoors. The quantum yield for the process was reported [29] to be 0.015 Einstein^{-1}:

$$-\!\!+\!CH_2\!-\!\!\underset{\underset{Cl}{|}}{CH}\!\!+_n \quad\longrightarrow\quad -\!\!+\!CH_2\!-\!\!\underset{\underset{Cl}{|}}{CH}\!\!+_{n-1}\!\!CH\!=\!CH\!-\!\!- \quad +\, HCl.$$

Photooxidation of PVC yielding carbonyl functionalities in the polymer is linear in time up to several hundreds of hours of exposure to xenon radiation. Using the cut-on filter technique and assuming the rate of carbonyl formation to be a linear function of the intensity of radiation, Martin and Tilley [82] obtained an activation spectrum for photooxidation of melt-pressed PVC films exposed to a xenon source. Maximum rate of degradation was observed at a wavelength of about 370 nm.

Solvent cast, oriented PVC films exposed to monochromatic radiation also underwent dehydrochlorination when exposed to radiation in the wavelength range of $\lambda > 340$ nm. The optical density of polymer changes as conjugated polyene sequences are produced by photodehydrochlorination. Action spectrum of the change in optical density per incident photon showed [83] a logarithmic dependence on irradiation wavelength for $\lambda > 254$ nm.

More relevant from a practical standpoint is the wavelength sensitivity of typical formulations (or compounds) of PVC after they have been processed. The most recently reported action spectrum of the polymer is for a rigid vinyl profile compound typically used in extruded siding and window frame applications. These compounds contain chalk, acrylic impact modifier, processing aid and an organo-tin mercaptide, in addition to the vinyl polymer, and were extruded through a ribbon die at about 360 °F. Andrady et al. [35] reported the action spectrum for photoyellowing of these compounds, with and without rutile titania opacifier, generated using near-monochromatic radiation in the spectral region $\lambda = 280$–400 nm. Rutile absorbs UV radiation and protects the formulations used outdoors from photodegradation. The Ln(Δ Yellowness Index (Y.I) per incident photon) varies linearly with wavelength of irradiation at 0, 2.5, and 5 phr of rutile titania in the formulation [35]. The action spectra for these samples were reported as follows:

$$\text{Ln}(\Delta \text{ Y.I/photon}) = a + b\lambda(\text{nm}). \tag{19}$$

Formulations with 0.0 phr of TiO_2 a = -31.1 b = -0.048 r = 0.99

2.5 phr or TiO_2 a = -30.6 b = -0.058 r = 0.98

5.0 phr of TiO_2 a = -27.9 b = -0.073 r = 0.99

The numerical value of the gradient b (nm^{-1}) quantifies the dependence of yellowing on the wavelength of irradiation.

Comparable data from exposures to equal numbers of incident photons (3×10^{19} photons/sq cm) at each wavelength agreed well with the above action spectrum [37], suggesting yellowing to be a linear function of intensity in the range of intensities encountered with the source used. The values of regression coefficients a and b given below were calculated from plots of Ln(ΔY.I. per 3×10^{19} photons/sq cm) vs the wavelength of exposure; data averaged from four replicate experiments yielded the following:

formulations with 0 phr of TiO_2 a = -11.96 b = -0.035 r = 0.95

Standard error of gradient 0.0025, n = 4.

(20)

The value of gradient obtained is close to the −0.048 obtained in the constant-duration exposure for the comparable formulation with no titania.

Wavelengths of 500 and 600 nm caused photobleaching rather than photo-yellowing of the polymer. The chemistry of this phenomenon, also reported by others [29], is not fully understood. Quenching of polyenyl radicals by oxygen [84, 85] and possible reaction of polyenes with hydrogen chloride formed during photodehydrochlorination, have been proposed [86, 87] as possible mechanisms.

Change in the stiffness of plasticized PVC (containing a ketone rubber plasticizer) as a function of the wavelength of irradiation has been reported [88]. Wavelengths below 360 nm had a marked stiffening effect and longer wavelengths up to 390 nm had a relatively smaller effect. Based on their data, the change in stiffness per available unit of irradiance can be calculated for five discrete wavelengths in the region of 330–410 nm. These damage estimates varied logarithmically with the wavelength of irradiation with a gradient of only −0.02 (nm^{-1}) $(r = 0.83)$.

Activation spectrum for yellowing of the same PVC compositions on exposure to a filtered xenon source, determined by the cut-on filter technique, has also been reported. The wavelength interval $\lambda = 310$–325 nm was identified as being the most effective in bringing about photoyellowing [37] of extruded PVC sheets with no titania and with 2.5 phr of titania. These findings are consistent with the conclusions from an earlier report of an activation spectrum derived from a monochromatic exposure experiment. For a similar source, the maximum degradation as detected by UV-visible spectral absorbance was found to be at $\lambda = 320$ nm [19].

5.2 Polycarbonate

Irradation of polycarbonates with UV-visible radiation can lead to extensive yellowing of the material. At least two different types of reactions responsible for yellowing have been postulated. When irradiated with short wavelength UV-B or UV-C radiation, the polymer undergoes photo-Fries rearrangement. Yellowing has often been attributed to this reaction and its products [89] such as o-dihydroxybenzophenones [90]. However, the reaction mostly occurs in the absence of, or at low levels of, oxygen. When irradiated at longer wavelengths (including solar visible wavelengths) [91], in the presence of oxygen, oxidative reactions are also believed to occur concurrently. These also result in the formation of colored products [92–94]. Some typical products from photodegradation of Bisphenol A polycarbonate are shown below. While quinones and salicylates are believed to be among the reaction products [94], the specific compounds responsible for yellow coloration have not been fully identified [95]:

Photo-Fries reaction

Oxidation products

Exposure of bisphenol A polycarbonate to near-monochromatic radiation of wavelengths 300, 280 and 260 nm results in chain scission of the polymer as evidenced by the decrease in solution viscosity [96]. Torikai et al. reported the wavelength-sensitivity of the quantum yield of chain scission as well as the estimated efficiency of photo-Fries reaction, assuming the absorption band at $\lambda = 320$ nm is assignable to the relevant reaction products [96]. The quantum yield spectrum for chain scission is shown in Fig. 2. Chain scission occurs via photolysis of phenyl ester linkage and subsequent decarbonylation of the primary radical to yield either a substituted phenoxy end group or a substituted phenyl radical at the point of chain-scission [97, 98]. The peak at about 325 nm in the absorption spectrum of yellowed polycarbonate has been assigned to the phenyl salicylate group [99], a product of photo-Fries reaction. Andrady et al. [100], however, suggest that the peak might be attributed to the formation of stilbene structures [101]. While some claim no photo-Fries reaction products to be formed by radiation of $\lambda > 310$ nm [102] or even by wavelengths as long as $\lambda = 365$ nm [92], others claim the lower limit to be close to $\lambda = 365$ nm [93, 95]. Resolution of this issue is important to the understanding of the role of photo-Fries processes initiated by terrestrial sunlight. Recently, Factor et al. [103], using mass spectrometric methods and radiolabelled polycarbonates, were able to show conclusively that natural weathering exposure yielded photo-Fries products.

Fukuda and Osawa [104] studied the photodegradation of polycarbonate films in air using the IR absorbance band at 1687 cm^{-1} to monitor the photo-Fries reaction products. Only the short wavelengths of $\lambda = 260$ and 280 nm showed any product formation; longer wavelengths yielded no measurable increase in the absorbance of irradiated samples. Absorbance in the UV (at 320 nm) also showed wavelength dependence in the interval of $\lambda = 260-320$ nm. Andrady et al. [105] reported the action spectrum for yellowing of commercial polycarbonate sheets by monochromatic radiation. These samples were 1.05 mm thick and contained a light absorber incorporated as a photostabilizer. The Ln (change in yellowness per incident photon) was shown to be a linear function of the wavelength of exposure. The two previous studies were also based on exposure to monochromatic radiation, and reported changes in the quantum efficiency of scission, Φ_S, deduced from viscosity [96] and the change in UV absorbance [91], with the wavelength of irradiation. The values of Ln (Φ_S) and Ln (Δ Absorbance) can be estimated from these published data and their linearity with respect to the wavelength of irradiation can be tested.

All three studies discussed above were carried out under similar conditions using the same source at Okazaki (Japan) but on different grades of bisphenol A polycarbonate. Figure 10 shows a comparison of the data. Given below are the gradients obtained from these plots of Ln (damage per photon) vs wavelength.

Reference	Property	Gradient	Value r^2
Andrady et al. 1992	Yellowness Index [100]	-0.082	0.99
Torikai et al. 1993	Quantum Yield (Scission) [96]	-0.044	0.99
Osawa et al. 1991	Absorbance (320 nm) [104]	-0.059	0.88

These data illustrate the dependence of action spectra on the type of photoreaction studied. The apparent higher wavelength sensitivity of the Yellowness Index might be due to the fact that, unlike the other two properties, it is a surface change. The high degree of linearity observed is significant and shows the effectiveness of photons in bringing about the photodamage to be logarithmically related to photon energy.

Activation spectra obtained using the filter technique have been reported for yellowing of bisphenol A polycarbonate sheets on exposure to a xenon source as well as to sunlight [100] (See Fig. 11). The maximum effect was obtained in the wavelength interval of $\lambda = 317-338$ nm in the case of exposure to solar radiation. A similar spectrum obtained using a borosilicate-filtered xenon source showed a maximum at $\lambda = 310-330$ nm, as might be expected due to the similarity in spectral irradiance distribution of the sources [100]. However, the latter also shows a maximum in the short wavelength interval of $\lambda = 272-293$ nm. This is due to the presence of small amounts of very short wavelength UV radiation in the xenon source spectrum, compared to terrestrial solar radiation. It is reasonable to conclude that the nature of yellowing is essentially the same under both exposures, and it is tempting to assign the short-wavelength peak observed with the xenon source exclusively to additional photo-Fries activity due to the shorter wavelengths in the source. An activation spectrum for the same polymer but for change in absorbance at 340 nm on exposure to a filtered xenon source was also reported [100]. Somewhat unexpectedly, the general features of the spectrum were similar and the maximum the

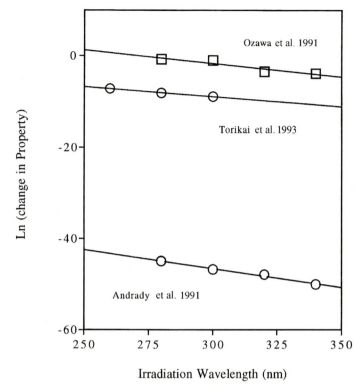

Fig. 10. Action spectra for photodegradation of polycarbonate at ambient temperature. Change in yellowness index from [105]; change in absorbance from [104]; change in quantum yield for chain scission from [96]

same as that for yellowing. An activation spectrum for yellowing of the same material obtained using a carbon-arc Fade-Ometer source[2] had quite different features and the maximum damage was observed at about $\lambda = 335-350$ nm. This difference is anticipated in view of the differences in the spectral irradiance distribution of the two sources.

Mullens and Searle [106] reported changes in absorbance of polycarbonate films (both solvent cast and extruded as 10-mil sheets) using a xenon source spectrum dispersed by a prism. The activation spectrum generated using this spectrographic technique showed maximum effectiveness in increasing the UV absorbance of films at around 330 nm in agreement with the reported data obtained with the cut-on filter technique.

5.3 Poly(methyl methacrylate)

High molecular weight PMMA ($M_V = 637\,000$) films cast from acetone and vacuum-dried were exposed to near-monochromatic radiation in vacuo to

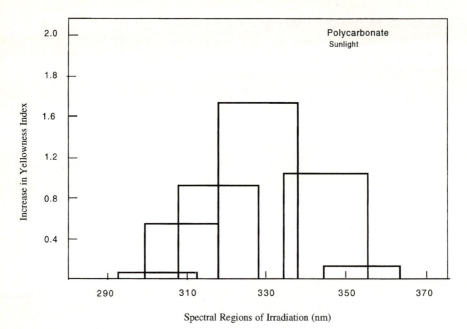

Fig. 11. Activation spectrum for the change in yellowness index of unstabilized polycarbonate (0.70-mm) exposed to sunlight (Miami, FL), at ambient temperature, determined by the cut-on filter technique [100]

obtain an action spectrum for chain scission [13]. Under UV irradiation, homolytic main chain scission and a secondary chain scission via an indirect mechanism resulting from ester side chain scission are known to occur [107, 108]:

Both reactions yield an unsaturated chain end and a propagating radical. The quantum yield for scission, Φ_S, was determined from viscometric data on irradiated films and known photon fluence levels, using an expression similar to the one below:

$$\Phi_{CS} = (N/M_n)\{(M_{V,o}/M_V - 1)/\{I_o(1 - 10^{-(\Sigma\varepsilon_i(\lambda)c_i)d}) \cdot t\} \tag{21}$$

where, $M_{V,o}$ and M_V are viscosity-average molecular weights at time 0 and at time, t, respectively, N is the Avegadro's number, d the thickness of the film, and M_n the number average molecular weight of polymer before irradiation. Total photon fluence was 4.21×10^{19} photons/sq cm. In this and other studies which rely on solution viscosity to determine the Φ_{CS}, it is important to recognize the assumptions implicitly made in the calculations. The substitution of $(\eta)^{1/a}$ in place of M_n assumes that the molecular weight distribution of the sample is not affected by the photodegradation. Furthermore, the time dependence of the extinction coefficient of chromophores, or the effects of the possible presence of microgel on viscometric data, are also not taken into account. While errors involved are expected to be small, their magnitudes have not been established. Figure 9 shows the action spectrum so obtained, with the most efficient scission obtained at $\lambda = 300$ nm wavelength of irradiation. The extent of chain scission was too small to be detected at irradiation wavelengths of 400 and 500 nm. Irradiation of PMMA in air under similar conditions [109] yielded higher quantum yields for scission at 260, 280 and 300 nm, with 300 nm radiation being the most efficient. An interesting observation in this work is the linear dependence of chain scission on the intensity of radiation at the same wavelength, and total photon fluence. Lower photon fluence rates were relatively more efficient in chain scission on a per photon basis than higher photon fluence rates within a narrow range of 2.7×10^{15}–4.7×10^{15} photons/sq cm. In a free-radical mediated chain scission occurring in the solid phase, increased radical-radical terminations at higher intensities may affect the efficiency of degradation. It is, however, interesting to establish if this linear dependence holds over a wider range of photon fluence.

The same group [110] has recently investigated the photodegradation of PMMA films containing a benzophenone sensitizer in low concentrations (0.05, 0.10 and 0.23 mol%). Gravimetrically determined gel fraction in samples exposed to equal photon fluences showed marked wavelength dependence with a maximum at approximately $\lambda = 340$ nm. This maximum was independent of the concentration of sensitizer but the amount of gel formation increased with its concentration in the polymer.

Fukuda and Osawa [104] using both monochromatic radiation exposures in the same wavelength range as above, and a high pressure mercury vapor lamp source, studied the change in absorbance of PMMA at $\lambda = 280$ nm on irradiation. In the wavelength range of 260–500 nm, no measurable effect was found.

5.4 Polystyrene

Polystyrene also undergoes photoyellowing on exposure to solar or xenon-source radiation [111–112]. The yellowing process is faster in nitrogen than in air [113] and was attributed to the formation of conjugated polyene sequences from terminally unsaturated fragments. Alternative hypotheses for yellowing due to formation of oxygenated species [114], or due to products of ring-opening reactions [115], have also been proposed. In spite of the large number of proposed candidates [116–119], the chromophores responsible for yellow coloration have not been unambiguously identified. The mechanical properties of polystyrene are also drastically affected by exposure to UV radiation [120, 122].

Three recent studies on the spectral sensitivity of polystyrene have been reported. An action spectrum of the change in optical density, as well as the number of chain scissions, of melt pressed films of polystyrene have been reported [121]. The change in optical density (at 310 nm) per incident photon was found to be a monotonically decreasing function of the wavelength of irradiation over the interval of $\lambda = 260–320$ nm. At a total fluence of 9×10^{19} photons/sq cm, no effect was obtained on irradiation at $\lambda = 320, 340,$ or 360 nm. The intensity of the carbonyl absorption peak in the infrared spectrum (1732 cm^{-1}) also increased on irradiation with 260 nm radiation. Chain scission events based on the number average molecular weight of the photodegraded material estimated using gel permeation chromatography was also reported. The technique, however, tends to underestimate the lower molecular weight fragments and does not take into account the possible presence of low levels of gel in the sample. Irradiation at 280 nm yielded the most efficient chain scission process. For polystyrenes containing flame retardants, either DBDE (decabromodiphenyl oxide) or TBA (tetrabromobisphenol A) at a 2% level, the maximum efficiency of chain scission was obtained at 300 nm. The presence of these flame retardants in the polymer also shifted the maximum of the action spectrum for the absorbance (at 310 nm) to longer irradiation wavelengths of $\lambda = 280–300$ nm.

Andrady [122] reported an activation spectrum for yellowing of expanded, extruded polystyrene sheets, exposed to filtered xenon-source radiation, using the cut-on filter technique. Yellowness Index as well as the L, a, b color parameters were determined. The activation spectrum for yellowing shown in Fig. 12 shows the most effective radiation band in the source spectrum to be $\lambda = 310–345$ nm.

As polystyrene substrates have a very high degree of whiteness it is particularly suitable for studying the change in hue of the sample as a function of the spectral irradiance distribution. A hue such as "yellow" is uniquely identified by a point in the L, a, b color space. The displacement of this point as a function of the cut-on filter used to modify the source spectrum indicates the progress of discoloration. Therefore, when changes in hue of the irradiated samples in "a − b" color space is plotted for samples exposed behind each filter, the general trend of the process becomes apparent. With polystyrene, it is clear that the

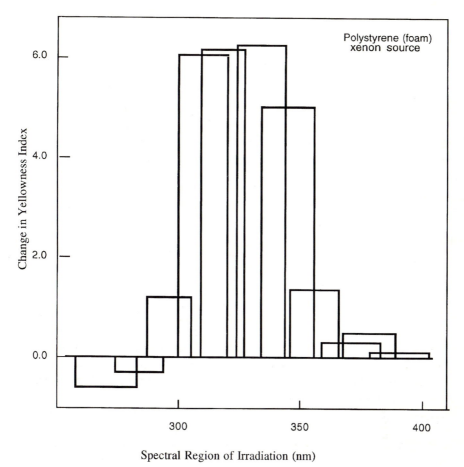

Fig. 12. Activation spectrum for change in yellowness index of extruded, expanded, polystyrene foam on exposure to a filtered xenon source (189 h exposure at 60 °C) determined by the cut-on filter technique [122]

samples exposed to longer wavelengths were white (unchanged from unexposed condition) with (a, b) close to zero. As progressively shorter wavelengths are introduced into the spectrum, the hue becomes increasingly yellow with a slight bias towards green. The data might be compared with that for processed PVC samples containing no titania opacifier, where the starting substrate itself is slightly green-yellow. On irradiation with progressively shorter wavelengths it yields increasingly yellow-red chromophores. With newsprint paper (ligno-cellulose), the change in parameter "a" with wavelength of irradiation is small, and yellow chromophores are developed with increasing levels of UV radiation reaching the sample during irradiation. These plots can be described by approx-imately linear functions in the a − b space (see Fig. 13). While the photon fluence afforded by different cut-on filters will be different, the three sets of data in the figure are comparable as they were generated using the same set of cut-on filters.

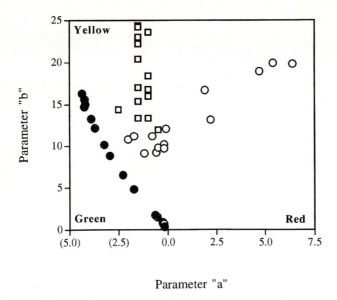

Fig. 13. Change in hue in the a–b color space, on exposure of sample to increasing amounts of UV radiation in the xenon source spectrum, by exposing behind different cut-on filters. ● expanded extruded polystyrene, ○ rigid PVC formulation with no titania, and ■ groundwood pulp

Spectral sensitivity of polystyrene to yellowing on exposure to white light (already referred to in Sect. 2.3) was reported by Allan et al. [20]. While no activation spectrum was reported, they obtained a quantum yield spectrum for yellowing of polystyrene.

5.5 Polyolefins

In spite of the numerous studies reported on photooxidation of polyolefins, the detailed mechanism of the complete process remains unresolved. The relative contribution by species involved in photoinitiation, the origins of the oxidative scission reaction, and the role played by morphology in the case of photoreactions in solid state are not completely understood. Primary initiator species in polyethylenes [123] and polypropylenes [124] are believed to be mainly ketones and hydroperoxides. During early oxidation hydroperoxides are the dominant initiator, particularly in polypropylene, and can be photolyzed by wavelengths in solar radiation [125]. Macro-oxy radicals from photolysis of polyethylene hydroperoxides undergo rapid conversion to nonradical oxy products as evidenced by ESR studies [126]. Some of the products formed are ketones susceptible to Norrish I and II reactions leading to chain scission [127, 128]. Norrish II reactions predominate under ambient conditions [129]. Concurrent with chain scission, crosslinking, for instance via alkoxy macroradical combination [126], can take place with consequent gel formation [130, 131].

5.5.1 Polypropylene

Exposure of polyolefins to UV radiation in the presence of air leads to autoxidative reactions with the incorporation of oxygenated groups into the polymer. The increase in UV absorbance (measured at 310 nm) of irradiated polypropylene is reported to be too low to measure for wavelengths of irradiation in the range of 260–600 nm. Even at 260 nm, only a minimal increase was observed [121]. For polypropylene containing flame retardants, either DBDE (decabromodiphenyl oxide) or TBA (tetrabromobisphenol A) at a 2% level, the maximum efficiency of chain scission was observed at $\lambda = 280$ and 260 nm, respectively.

Mechanical properties of polyolefins deteriorate on exposure to UV radiation [130, 132]. A cumulative spectral sensitivity curve for changes in tensile properties of injection-molded unstabilized polypropylene laminates (68-mil) exposed to a filtered xenon source radiation, generated using a series of cut-on filters, is available [133]. The wavelength interval of $\lambda = 320–360$ nm was found to be the most effective in reducing tensile strength of the polypropylene samples. A second set of data generated for a different grade of polypropylene similarly processed and exposed showed the maximum loss in tensile strength to occur at $\lambda = 315–325$ nm; the most effective waveband in terms of reduction in tensile yield stress was $\lambda = 315–330$ nm. Tensile properties of polypropylene showed a high degree of variance in general and the high variance in elongation at break values precluded its use to obtain an activation spectrum.

Same material but containing 0.1% of a photostabilizer (Cyasorb UV 3346, American Cyanamid Co.) was also studied in a similar experiment designed to obtain a direct activation spectrum of the material [133]. After 1400 h exposure in an Atlas Ci 65 weatherometer, the samples showed measurable changes in both elongation at break as well as the tensile strength. Maximum effectiveness was found at $\lambda = 335–355$ nm.

Recently, Zhengfeng reported [134] an activation spectrum for changes in optical density due to carbonyl functionalities, and in tensile strength, for polypropylene exposed to both filtered xenon radiation and to sunlight (Guangzhou, China). They found qualitative similarities between the two exposures and maximum degradation in the interval $\lambda = 300–350$ nm in several exposures spanning durations of up to 136 h.

5.5.2 Polyethylenes

Polyethylene, the most-used commodity thermoplastic, also shows wavelength-dependent behavior in photodegradation. Heacock [135] proposed UV radiation of about $\lambda = 257$ nm to result in the formation of unsaturated and oxygenated structures in polyethylene, while the longer wavelength UV-A radiation was believed to cause mainly crosslinking reactions. Results from a monochromatic exposure study of low-density and high-density polyethylene films were recently reported [136]. At a total photon fluence of 8×10^{19} photons/sq cm, a significant increase in the absorbance of the films (measured at 310 nm) was obtained with irradiation at 260 nm; longer wavelengths investigated from $\lambda = 290$–360 nm did not cause a measurable change in absorbance. The presence of flame retardants, DBDE (decabromodiphenyl oxide) or TBA (tetrabromobisphenol), in the film resulted in a higher level of UV absorbance but the origin of the absorbance itself is not known. It is difficult to interpret the data because of the uncertain role of the flame retardant in the photoreactions and the possibility of at least part of the absorbance being attributable to possible photodegradation products of the retardant itself.

A cumulative spectral sensitivity curve for the change in tensile properties of injection molded samples (68-mil) of unstabilized low density polyethylene exposed to a xenon source has also been carried out [133]. Both the mean elongation at break of the samples, as well as the mean tensile strength of samples, were measured; the spectral region of maximum effectiveness was identified as $\lambda = 308$–329 nm (see Fig. 14). A set of similar samples, but containing 0.05% of Cyasorb 3346 light stabilizer, exposed over 5000 h to a xenon source yielded a similar cumulative spectral sensitivity curve. The region of maximum change was shifted to slightly shorter wavelengths of $\lambda = 286$–304 nm.

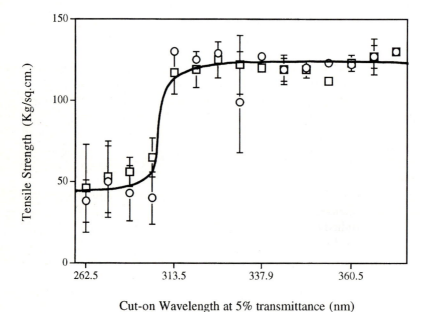

Fig. 14. A cumulative spectral sensitivity curve for the decrease in average elongation at break of injection molded unstabilized LDPE samples exposed to filtered-xenon radiation, behind a series of different cut-on filters. Two sets of data for different commercial LDPE resins are shown. The *error bars* show the standard error of the mean [133]

The above studies were tedious because of the very long exposure times required, especially in the case of photostabilized samples, to achieve a significant change in tensile properties. Shorter exposure durations remove some of the experimental difficulties (such as the need to rotate samples within an exposure chamber) and the cost of the experiment. Using thin films instead of thick samples and a catalyst to accelerate photodegradation of polyethylene, might shorten the exposure time. An LDPE film containing less than 100 ppm of an iron compound prooxidant was used in a cut-on filter exposure to determine the cumulative spectral sensitivity curve for the decrease in tensile properties on exposure to a xenon source spectrum [122]. A fairly sharp change in the mean tensile elongation at break was observed in the wavelength interval of $\lambda = 340$–352 nm. The maximum is at a somewhat higher wavelength compared to that obtained for tensile strength of injection molded polyethylene. This might be due to catalysis of thermal oxidation by the metal compound prooxidant affecting the wavelength sensitivity of the chain scission process.

Trubiroha [43] reported an action spectrum for development of carbonyl groups (as measured by the change in absorbance at 5.8 microns) of unstabilized polyethylene and polypropylene films. A measurable effect after 1 MJ/sq m of radiant exposure was obtained only for $\lambda = 380$ nm; between 280 and 380 nm the absorption changed logarithmically with wavelength of irradiation.

5.6 Polyamides

The change in average molecular mass of nylon 6 (39-micron film) and nylon 66 (31-micron film) on exposure to monochromatic radiation in the wavelength range of $\lambda = 280$–380 nm has been reported. The fractional viscosity-average molecular weight of the irradiated samples (1 MJ/sq m) decreased logarithmically with decreasing wavelength of irradiation. The threshold wavelengths observed at the specified film thickness were $\lambda = 340$ nm for nylon 66 and $\lambda = 370$ nm for nylon 6 samples.

The high temperature resistant polyamides, a poly(isophthalamide) of 4,4'-bis(p-aminophenyl)-2,2'-bithiazole) (TBA), a poly-terephthalamide of 1,3-bis(3-aminobenzamido)benzene (M3P), and the aromatic polyamide Nomex (DuPont), were studied by Johnson et al. [31]. Activation spectra for discoloration and strength loss were obtained using a prism-dispersed spectrum from a xenon source using a technique similar to that used by Hirt and Searle [19].

The TBA was discolored most efficiently by the $\lambda = 370$–500 nm band of the source spectrum. The M3P and Nomex samples showed maxima at 370 nm and 360 nm, respectively, at shorter wavelengths relative to that for TBA. Losses in strength of the polyamide fibers M3P and Nomex were also measured, and the maximum efficiency of the processes was attributed to the same waveband as that for discoloration.

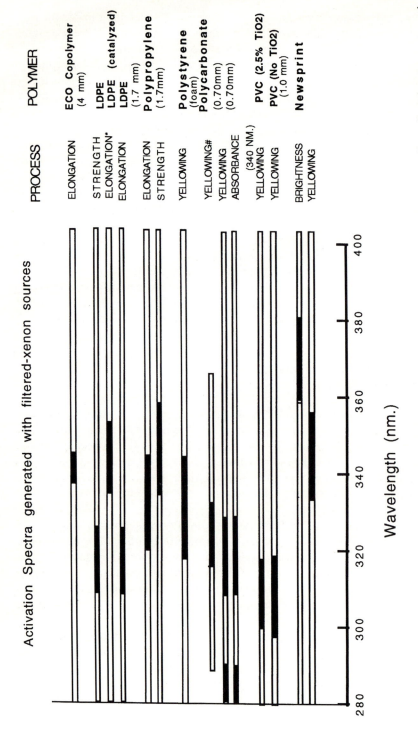

Fig. 15. A summary of reported activation spectra for polymers, for change in different properties on exposure to a borosilicate-filtered xenon source, and determined by the cut-on filter technique

6 Conclusions

A large number of photodegradation processes commonly met with polymers show some degree of wavelength sensitivity. These might be expressed in terms of the effectiveness of incident photons in yielding a given photoreaction as a function of the wavelength of irradiation to obtain an action spectrum. The spectral sensitivity of a process to various spectral regions or wavebands of a selected source spectrum is also of interest and is referred to as an activation spectrum. However, it is possible to derive either spectrum from experimental data pertaining to the other. Thus an action spectrum can be calculated from an activation spectrum and vice versa. Such calculated spectra need not always be identical to the direct measurements, due to several reasons. The most important of these are that the degradation may not be a linear function of the dose of radiant energy, and that any synergistic or antagonistic effects by different wavelengths is precluded in monochromatic exposure experiments but permitted in polychromatic exposures. Even when these conditions are met the differences in experimental techniques need to be taken into account, and, at best, only agreement in trends of data might be expected.

Action spectra determined by exposure of samples to near-monochromatic radiation and subsequent measurement of properties of interest have been reported for a number of samples. As a consequence of the functional form of the dependence of photon energies on wavelength, plots of damage per constant photons vs wavelength tend to be logarithmic. Also reported in the literature are activation spectra which refer to the effectiveness of different regions of a specific source spectrum in causing a photoreaction. Reported activation spectra determined using the cut-on filter technique and a borosilicate-filtered xenon source (usually a Ci-65 weather-Ometer) are shown in Fig. 15. Depending on the photoprocess of interest, the wavelength interval of maximum effectiveness might be located in the UV-B or UV-A region of the white-light spectrum.

The review illustrates the limited amount of data on wavelength sensitivity available in the literature. This is particularly true of polymer formulations of practical interest including compounded and processed polymers. As the technique, especially in the case of activation spectra, is tedious, often no error limits for the spectra have been established. Furthermore, the assumption that wavelength sensitivity data is time-independent is implicit in most published data; in most instances not enough data is available to test the validity of this assumption.

Acknowledgements. The author wishes to acknowledge the support of Research Triangle Institute in researching this review. The author is especially grateful to Dr. Norma Searle (Deerfield Beach, FL) for the numerous helpful suggestions to improve this manuscript. The suggestions made by Dr. Norbert Maecker (Dow Chemical Company, Department of Designed Thermoplastics Research, Midland, MI) Professor Ayako Torikai, Nagoya University (Japan), and Professor Vivian Stannett (Research Triangle Institute), are also thankfully acknowledged.

7 References

1. Stivala SS, Kimura J, Gabbay SM (1993). In: Allen NS (ed) Degradation and stabilization of polymers, Applied Science Publishers, New York, p 79
2. Searle ND (1989). In: Patsiis AV (ed) International conference on the advances in stabilization and controlled degradation of polymers, Vol I. Technomic Publishing, Lancaster Pa, p 62
3. Phillips D (1985). In: Phillips D (ed) Polymer photophysics. Chapman and Hall, New York
4. Turro N (1978) Modern molecular photochemistry. Benjamin/Cummings, Menlo, Park, CA
5. Billingham NC, Walker TJ (1975) J Polym Sci A1 13: 1209
6. Rabek F (1995) Polymer photodegradation, Chapman and Hall (London, England)
7. Rabek JF (1987) Mechanisms of photophysical processes and photochemical reactions in polymers, John Wiley and Sons (London)
8. Shlyapintokh VY (1985) Photochemical conversions and polymer stabilization, Hanser Publishers, Munchen, Germany
9. Scott G (1990) Polymer degradation and stabilization, Elsevier Applied Science Publishers (London)
10. Osawa Z (1986) Photodegradation and stabilization of polymers: Fundamentals and practice of photostabilization of polymers, CMC Company (Tokyo, Japan)
11. McKellar JF, Allen NS (1979) Photochemistry of man-made polymers, Applied Science Publishers (London, England)
12. Schafer E, Fukshansky L, Shropshire W Jr (1983) In: Shropshire W Jr, Mohr H (ed) Encyclopedia of plant physiology (Springer Verlag, Berlin)
13. Decker C, Balandier M (1982) Eur Polym J 18: 1085
14. Gibb WH, MacCuallum JR (1971) Europ Polym J 7: 1231
15. Feedasova GT, Strelkova LD, Krats EO, Lebedev VP, Wall KS1LA (1960) SPE J 16: 1031
16. Mitsuoka T, Torikai A, Fueki K (1993) J Appl Polym Sci 47: 1027
17. Balandier M, Decker C (1978) Europ Polym J 14: 995
18. Martin JW, Lechner JA, Varner RN (1994). In: Warren D. Ketola, Douglas Grossman (eds) Accelerated and Outdoor Durability Testing of Organic Materials, ASTM STP 1202, American Society for Testing and Materials, Philadelphia, p 27
19. Hirt RC, Searle NZ (1967) Appl Polym Symp 4: 61
20. Allan DS, Maecker NL, Priddy DB, Schrock NJ (1994) Macromolecules 27 (26): 7621
21. Dobrowolski JA (1965) Appl Opt 4: 937
22. Walraven RL (1978) Rev Sci Instrum 49: 537
23. Bradley DJ, Mitchell CJ (1968) Phil trans Royal Soc A 263: 209
24. Decker C (1984) Eur Polym J 20: 149
25. Decker C, Ackhardt A, Ehrburger P (1990) Carbon 28: 246
26. Garrison LM, Murray LE, Doda DD, Green AES (1978) Appl Opt 17: 827
27. Mertz L (1977) Appl Opt, 16: 3122
28. Watanabe W, Furuyu M, Miyoshi Y, Inoue Y, Iwahashi I, Matsumoto K (1983) Bioscience 33: 53
29. Decker C, Balandier M (1981) J Photochem 15: 221
30. Pochan JM, Gibson HW, Baily FC (1980) J Polym Sci, Polym Lett Ed, 18: 447
31. Johnson LD, Tincher WC, Bach HC (1969) J Appl Polym Sci 13: 1825
32. Searle ND, Preprint 7th International Conference on the Advances in Stabilization and controlled degradation of polymers, Luzern 22–24 May, 1985
33. Trubiroha P (1980) Materialpruf 22(3): 117
34. Davis A, Gordon D (1974) J Appl Polym Sci 18: 1173
35. Andrady AL, Torikai A, Fueki K (1989) J Appl Polym Sci 37: 935
36. Andrady AL, Fueki K, Torikai A (1990) J Appl Polym Sci 39: 766
37. Andrady AL, Searle ND (1989) J Appl Polym Sci 37: 2789
38. Shultz AR, Andrady AL (1987) J Appl Polym Sci 33: 2249
39. Andrady AL, Shultz AR (1987) J Appl Polym Sci 33: 1389
40. Kubelka P, Munk F (1931) Z Tech Phys 12: 593
41. Seyfried M, Fukshansky L (1983) Appl Optics 22(9): 1402
42. Andrady AL, Searle ND, Unpublished data.
43. Trubiroha P (1989). In: Patsis AV (ed) International conference on the advances in stabilization and controlled degradation of polymers, Vol I. Technomic Publishing, Lancaster Pa, p 236
44. Andrady AL, Searle ND (1995) TAPPI Journal 78(5): 131

45. Timell TE (1967) Wood Sci Technol 16: 83
46. Hon DN-S, Glasser W (1979) Plast Technol Eng 12: 159
47. Norrstrom H (1969) Sven Papperstidn 72: 25
48. Heitner C (1993). In: Heitner C, Sciano JC (eds) Photochemistry of lignocellulosic materials, Acs Symposium Ser 531 American Chemical Society, p 2
49. Gellerstedt G, Pattersson E (1977) Swensk Papperstidning 1: 15
50. Lin Y, Kringstad KP (1971) Norsk Skogindustri 25: 252
51. Lin SY, Kringstad KP (1970) Tappi 53: 1676
52. Brunow G, Eriksson B (1977) Acta Chem Scand 25: 22
53. Kringstad KP, Lin SY (1970) Tappi 53: 2296
54. Gierer J, Lin SY (1972) Svensk Papperstidn 75: 233
55. Leary GJ (1968) Tappi 51: 257
56. Nolan P, Van der Akker JA, Wink WA (1945) Pap Trade J 11(121): 33
57. Van der Akker JA, Lewis HF, Jones GW, Buchanan MA (1949) Tappi 4(32): 187
58. Forsskåhl I, Tylli H (1993) In: Heitner C, Sciano JC (eds) Photochemistry of lignocellulosic materials, Acs Symposium Ser 531 American Chemical Society. p 45
59. Andrady AL, Song Y, Parthasarathy VR, Fueki K, Torikai A, TAPPI 74(8): 162
60. Maclaren JA (1963) Textile Res J 33: 773
61. Launer HF (1965) Textile Res J 35: 395
62. Derbyshire AN, Lemin DR (1964) J Soc Dyers and Colourists 80: 243
63. Nicholls CH, Pailthorpe MT (1976) J Textile Inst 67: 397
64. Tokuyama T, Senoh S, Sakan T, Brown KS, Witkop B (1967) J Amer Chem Soc 89: 1017
65. Holt LA, Milligan B (1973) Aust J Biol, 26: 871
66. Holt LA, Milligan B (1976) J Text Res 67: 269
67. Lennox FG, King MG (1968) Textile Res J 38 754
68. Lennox FG, King MG, Leaver IH, Ramsay GC, Savige We (1971) Applied Polym Symp No 18: 353
69. Stroud D, Vinyl J (1983) Technol 5(2): 52
70. Summers JW, Rabinovitch EB (1983) J Vinyl Technol 5(3): 91
71. Mathur K, Kramer K (1983) J Vinyl Technol 5(1): 32
72. Summers J (1983) J Vinyl Technol 5(2): 43
73. Marechal JC (1978) J Macromol Sci Chem 12: 609
74. Verdu J (1978) Macromol Sci Chem 12: 551
75. Balandier M, Decker C (1978) Eur Polym J 14: 995
76. Daniels VD, Rees HH (1974) J Polym Sci Polym Chem Ed 12: 2115
77. Owen ED, Bailey RJ (1972) J Polym Sci Polym Chem Ed 10: 113
78. Braun D, Kull S (1980) Angew Makromol Chem 85: 79
79. Braun D, Kull S (1980) Angew Makromol Chem 86: 171
80. Rabonavitch EB, Queensberry JG, Summers JW (1983) J Vinyl Technol 5(3): 110
81. Braun D, Bezdadea E (1981) Angew Makromol Chem 99: 55
82. Martin KG, Tilley RI (1971) Br Polym J 3: 36–40
83. Reinisch RF, Gloria HR, Wilson DE (1966) Polym Preprints 7(1): 372
84. Druesdow D, Gibbs C (1953) Nat Bur Stand (USA) Circ 525: 69
85. Nagy TG, Kelen T, Turcsanyi B, Tudos F (1977) J Polym Sci, Polym Chem Ed 15: 853
86. Owen ED, Williams JI (1974) J Polym Sci, Polym Chem Ed 12: 1933
87. Owen ED, Read RL (1979) J Polym Sci, Polym Chem Ed 17: 2719
88. Warner WC, Gruber EE (1966) I&EC Proc Res Dev 5(3): 219
89. Gupta A, Rembaum A, Moacanin J (1978) Macromolecules 11(6): 1285
90. Ong E, Bair HE (1979) Polym Prep 20: 945
91. Rivaton A, Sallet D, Lemaire J (1983) Polymer Photochem 3: 463
92. Rivaton A, Sallet D, Lemaire J (1986) Polymer Deg Stab 14: 1
93. Clark DT, Munro HS (1984) Polymer Deg Stab 8: 195
94. Factor A, Chu ML (1980) Polymer Deg Stab 2: 203
95. Clark DT, Munro HS (1982) Polymer Deg Stab 4: 441
96. Torikai A, Mitsuoka T, Fueki K (1993) J Polym Sci A, Chem Ed 31: 2785
97. Gesner BD, Kelleher PG (1969) J Appl Polym Sci 132183
98. Humphrey JS, Shultz AR, Jacquiss DBG (1973) Macromolecules 6: 305
99. Gupta VP, Pierre LE (1979) J Polym Sci, Polym Chem Ed 17: 797
100. Andrady AL, Searle ND, Crewdson LFE (1992) Polym Degrad Stab 35: 235–247

101. Factor A, Ligon WV, May RJ (1987) Macromolecules 20: 2461
102. Abbas K (1975) Appl Polym Symp 35: 345
103. Factor A, Ligon WV, May RJ Greenberg FH (1989). In: Patsis AV (ed) International conference on the advances in stabilization and controlled degradation of polymers, Vol II. Patsis Technomic Publishing, Lancaster Pa, p 45
104. Fukuda Y, Ozawa Z (1991) Polym Deg Stab 34: 75
105. Andrady AL, Fueki K, Torikai A (1991a) J Appl Polym Sci 42: 2105
106. Mullens PA, Searle NZ (1970) J Applied Polym Sci 14: 765
107. Gupta A, Lian R, Tasy FD, Mocanin J (1980) Macromolecules 13: 1696
108. Torikai A, Fueki K (1982) Polym Photochem 2: 297
109. Torikai A, Ohono M, Fueki K, (1990) J Appl Polym Sci 41: 1023
110. Torikai A, Hattori T, Eguchi T (1995) J Polym Sci, Polym Chem Ed 33: 1867
111. Andrady AL, Pegram JE (1991) J Appl Polym Sci 42(6): 1589
112. Gugumus F (1979) Dev Polymer Stab 1: 8
113. Grassie N, Weir NA (1965) J Appl Polym Sci 9: 987
114. Savides C, Stretanski JA, Costello LR (1968). "Light Stabilization of Polystyrene," in: Gould RF (ed) Stabilization of polymers and stabilizer processes, (Advances in Chemistry Series No. 85), American Chemical Society, Washington, DC, 1968, p 187
115. Ranby B, Lucki J (1980) Pure Appl Chem 52: 295
116. Lawrence B, Weir NA, (1966) Chem Commun 273
117. Kubica J, Waligora B (1977) Europ Polym J 13: 325
118. Rabek JF, Ransby B (1974) J Polym Sci, Polym Chem Ed 12: 273
119. Grassie N, Weir NA (1965) J Appl Polym Sci 9: 975
120. Ghaffar A, Scott GA, Scott G (1976) Eur Polym J 11: 271
121. Torikai A, Kato H, Fueki K, Suzuki Y, Okizaki A, Nagata M (1993) J Appl Polym Sci 50: 2185
122. Andrady AL, "Accelerated environmental exposure, Laboratory testing, and recyclability study of photo/biodegradable plastics", Final Report to USEPA under Contact # 68-02-4544 Task II-60. January 1991.
123. Scott G Ed, Ultraviolet light induced reactions in polymers. American Chemical Society, Washington, DC, 1976, p. 340
124. Carlesson DJ, Wiles DM (1976) J Macromol Sci Rev Macromol Chem C 14: 155
125. Benson W (1965) J Chem Educ 42: 501
126. Tsuji K, Nagita H (1977) Rep Prog Polym Phys Japan 20: 563
127. Davis W, Noyes WA (1947) J Amer Chem Soc 60: 2153
128. MacMillan GR, Clavert JG, Pitts JN (1964) J Amer Chem Soc 86: 3602
129. Guillet JE, Dhanaray J, Colemba FJ, Hartley GH (1968) Adv Chem Series 85: 272
130. Scott G (1976) J Polym Sci Symp 57: 357
131. Scott G (1975) Amer Chem Soc Div Org Coat Plast Chem, Prepr 35: 163
132. Omichi H, Hagiwara M, Asano M, Araki K (1979) J Polym Sci, Symp 24: 2311
133. Andrady AL, Pegram JE, Searle ND, Unpublished data
134. Zhengfeng Z, Xingzhou H, Zubo L, Unpublished data. Personal communication
135. Heacock JF (1968) J Polym Soc Part A-1 6: 2921
136. Torikai A, Chigata K, Okisaki F, Nagata M (1993) Polym Preprints Japan 42: 2045: J Appl Polym Sci (in press)

Editor: J.L. Koenis
Received: January 1996

Reaction-Induced Phase Separation
in Modified Thermosetting Polymers

Roberto J.J. Williams[1], Boris A. Rozenberg[2] and Jean–Pierre Pascault[3]
[1] Institute of Materials Science and Technology (INTEMA),
University of Mar del Plata and National Research Council (CONICET),
J.B. Justo 4302, (7600) Mar del Plata, Argentine
[2] Institute of Chemical Physics in Chernogolovka Russian Academy
of Sciences, Chernogolovka, Moscow region, 142432 Russia
[3] Laboratoire des Matériaux Macromoléculaires, URA CNRS No. 507,
Institut National des Sciences Appliquées de Lyon, 20,
Avenue A. Einstein, 69621 Villeurbanne Cedex, France

Thermosetting polymers are frequently used in formulations, including rubbers, thermoplastic polymers or oils, etc, in an amount of the order of 2–50 wt% with respect to the thermoset. This extra component, called the modifier, may initially be immiscible or may phase-separate during cure. This last process, i.e. the reaction-induced phase separation, is the subject of this review. A thermodynamic description of the process is made, using the Flory–Huggins equation at two approximation levels, i.e. a quasi-binary approach and a multicomponent treatment taking polydispersity of constituents into account. Thermodynamic factors affecting the phase separation process are thus established. Nucleation and growth (NG) and spinodal demixing (SD) are considered as possible phase separation mechanisms. Factors promoting one or the other process are discussed. The control of morphologies generated is analyzed on the basis of thermodynamic and kinetic arguments. Ideas for obtaining particular morphologies enhancing particular properties are put forward.

Advances in Polymer Science, Vol. 128
© Springer-Verlag Berlin Heidelberg 1997

List of Symbols and Abbreviations

a	parameter included in the relationship between the Flory–Huggins interaction parameter and the absolute temperature
A_4	tetrafunctional monomer
b	slope of the relationship between the Flory–Huggins interaction parameter and the inverse of absolute temperature
b_1, b_2	constants included in the dependence of χ with temperature and composition
B_2	difunctional monomer
C_{part}	concentration of dispersed phase particles
$\bar{D}$	average diameter of dispersed phase particles
d_0, d_1, d_2	constants included in the dependence of χ with temperature and composition
E	elastic (Young's) modulus
$E_{m,n}$	oligomer produced by polycondensation, containing n molecular units of monomer and m molecular units of comonomer (hardener)
g	interaction parameter depending on both temperature and composition
G	Gibbs free energy
g_0, g_1, g_{10}, g_{11}	constants included in the dependence of g with temperature and composition
G_{Ic}	fracture energy
h	parameter in the Schultz–Zimm equation
H	enthalpy
I_I	integrated intensity of the scattered light
$I_{(q)}$	intensity of the scattered light as a function of the scattering vector
K	ratio of the phase separation rate with respect to the cure reaction rate
k_B	Boltzmann constant
K_{Ic}	critical stress intensity factor
M	modifier
$\bar{M}_i$	molar mass of component i
$\bar{M}_n$	number-average molar mass
N	number of moles
p	conversion
P	average species of the polymer
P_i	macromolecule containing i monomer units

$P_{i,j,k}$	macromolecule obtained by chain polymerization, containing i monomer units, j functional groups and k active centres
q	scattering vector
r	ratio of amine to epoxy equivalents
R	gas constant
S	entropy
t	time
T	absolute temperature
T_g	glass transition temperature
$_{gel}T_g$	temperature where gelation and vitrification take plase simultaneously
Tg_∞	ultimate glass transition temperature of the polymer network
T_r	reaction temperature
U	parameter in the Schulz–Zimm equation
V	molar volume
v_0	initial reaction rate
V_D	volume fraction of dispersed phase
V_{MS}	molar volume of the unit segment of a polymeric modifier
V_{PS}	molar volume of the unit segment of the thermosetting polymer
V_r	molar volume of the unit cell (reference volume)
V_T	total volume of the system (extensive property)
x	denotes the x-mer of a polymeric modifier
$\overline{X_n}$	number-average degree of polymerization
$\overline{X_w}$	mass-average degree of polymerization
y	denotes the y-mer of the thermosetting polymer
Z_c	coordination number of a unit cell
Z_i	ratio of the molar volume of component i with respect to the molar volume of the unit cell

Greek letters

α-phase	thermoset-rich phase
α_i	thermal expansion coefficient of component i
β-phase	modifier-rich phase produced in the course of the primary phase separation
χ	Flory–Huggins interaction parameter
δ	solubility parameter
δ-phase	modifier-rich phase inside particles of the β-phase (secondary phase separation)
ΔG	Gibbs free energy of mixing per unit volume of system
ΔG_c	free energy barrier for nucleation
ΔG^M	Gibbs free energy of mixing (extensive property)

ΔH	enthalpy of mixing per unit volume of system
ΔH^M	enthalpy of mixing (extensive property)
ΔS	entropy of mixing per unit volume of system
ΔS^M	entropy of mixing (extensive property)
$\Delta \mu_i$	chemical potential of component i
ε'	exchange energy
ε_{ij}	energy of a contact between components i and j
ϕ	volume fraction
γ	constant included in the dependence of γ with temperature and composition
γ-phase	thermoset-rich phase inside particles of the β-phase (secondary phase separation)
Γ	gamma function
η	viscosity
Λ	interaction parameter with units of energy per unit volume
σ_0	interfacial tension
σ_y	yield stress
ω	mass fraction

Subscripts

CP	value of the cloud point
crit	value at the critical point
gel	value at the gel point
M	modifier
0	initial value
P	polymer
S	silicon chip
x	x-mer of a polymeric modifier
y	y-mer of the thermosetting polymer

Abbreviations

AN	acrylonitrile
ATBN	amino-terminated butadiene-acrylonitrile copolymer
BD	butadiene
BMC	bulk moulding compound
CE	cyanate ester
CO	castor oil
CP	cloud point
CPC	cloud-point curve
CS	core-shell particles
CTBN	carboxyl-terminated butadiene-acrylonitrile copolymer
3DCM	4,4'-diamino-3,3'-dimethyldicyclohexyl-methane
DGEBA	diglycidyl ether of bisphenol A

EDA	ethylenediamine
ETBN	epoxy-terminated butadiene-acrylonitrile copolymer
FH	Flory–Huggins
GMA	glycidyl methacrylate
HIPS	high-impact polystyrene
IPN	interpenetrated polymer network
LCST	lower-critical-solution temperature
LS	light scattering
LT	light transmission
NFBN	non-functionalized copolymer of butadiene and acrylonitrile
NG	nucleation-growth
NR	nucleation rate
PDLC	polymer-dispersed liquid crystal
PE	polyester
PEI	poly(etherimide)
PES	poly(ethersulfone)
PICS	pulse-induced-critical scattering
PVAc	poly(vinylacetate)
PVME	poly(vinylmethylether)
R	rubber
S	spinodal curve
SAXS	small-angle-X-ray scattering
SBR	styrene butadiene random copolymer
SD	spinodal demixing
SEM	scanning electron microscopy
SMC	sheet moulding compound
SZ	Schulz–Zimm equation
TEM	transmission electron microscopy
TP	thermoplastic
TTT	time-temperature-transformation diagram
UCST	upper-critical-solution temperature
UP	unsaturated polyester
UV	ultraviolet light

1 Introduction

1.1 Modified Thermosetting Polymers

Thermosetting polymers, like phenolics, epoxies, unsaturated polyesters, etc, are frequently used in formulations containing a low or high-molar-mass rubber, a thermoplastic polymer, an oil, etc, in an amount of the order of 2–50 wt% with respect to the thermoset. This extra component, called the modifier, may initially be immiscible or may phase-separate during cure (reaction-induced phase separation).

The modifier is added for different purposes.

a) Toughening of the polymer network. This means that values of fracture energy (G_{Ic}) or critical stress intensity factor (K_{Ic}) are increased through the generation of a second-phase dispersion with particles in the micrometer range. The situation is typical of rubber-modified epoxies used as coatings, structural adhesives or matrices of composite materials. Typical properties of rubber-modified epoxies are shown in Fig. 1 as a function of the rubber amount (5–20 wt%). The increase in G_{Ic} is due to several toughening mechanisms such as particle cavitation, matrix shear yielding and rubber bridging [1]. The introduction of a rubbery phase leads to a decrease in the elastic modulus (E) and yield stress (σ_y). The deterioration of these properties may be avoided if the rubber is replaced by an engineering thermoplastic like poly(etherimide) (PEI) or poly(ether sulfone) (PES), exhibiting similar values of elastic modulus and yield stress as those of the epoxy network.

Several books have been devoted to the toughening of thermoplastic and thermosetting polymers [2–4].

When the process involves two competitive reactions, some people prefer to call those modified polymers interpenetrated polymer networks (IPNs) [5]. The formation of a polyether-urethane network in a loosely crosslinked poly(methyl methacrylate) matrix to increase its toughness can serve as one of the examples. From a general point of view, the analysis of the reaction-induced phase separation is the same (perhaps more complex) for IPNs than for rubber-modified epoxies or for high-impact polystyrene.

b) Improving the quality of molding surfaces through the use of low-profile modifiers. The use of poly(vinyl acetate) (PVAc, 5–10 wt%) in an unsaturated polyester (UP) resin used in sheet or bulk moulding compound (SMC or BMC) formulations is a typical example. Phase separation of the low-profile modifier occurs at the very beginning of the chain polymerization [6, 7]. At low concentrations of the modifier, the morphology is characterized by discrete particles rich in PVAc dispersed in the cured UP matrix. At concentrations higher than the critical value, a bi-continuous structure is obtained [8, 9]. These bicontinuous regions promote cavitation in response to tensile stresses arising from

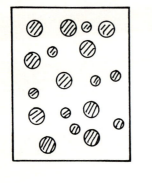 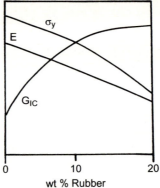

0   10   20
wt % Rubber

Fig. 1. Typical properties of rubber-modified epoxies as a function of the rubber amount (E = elastic modulus, σ_y = yield stress, G_{Ic} = fracture energy)

internal thermal and cure contractions, in combination with mechanical constraints [10]. Typical properties of PVAc-modified UP resins are shown in Fig. 2. The surface rugosity exhibits a sharp decrease beyond the critical PVAc concentration [11]. However, the impact resistance is significantly reduced due to the presence of a weak PVAc bicontinuous phase. Therefore, in this case the purpose of the modification, i.e. to produce an excellent copy of the mould surface, has a deleterious effect on mechanical properties.

c) Producing interconnected microporous structures in a coating based on a thermosetting polymer. A typical example is the use of an epoxy-phenolic matrix (polycondensation) modified with poly(vinyl methyl ether) (PVME) in magnetic disk coatings [12]. A second phase rich in PVME is segregated during the cure of the thermosetting polymer. Using a convenient percentage of PVME in the initial formulation, a bicontinuous structure may be generated. When heating up to 220 °C in air, oxidative degradation of PVME takes place leading to a crosslinked polyene structure and emission of volatile products. The morphology of the microporous structure thus generated depends on the initial reaction-induced phase separation of PVME. The presence of microporosity in the coating enhances the retention of a lubricant used to increase the durability of the head-disk interface.

d) Reducing thermal stresses through lowering of the Young's modulus of the polymeric material. This effect is of interest for epoxy resins used for electronic packaging. Internal stresses initiate package cracking and deformation. A differential shrinkage of the electronic device and the resin, arising from the difference of thermal expansion coefficients during the cooling cycle to room temperature, is the main source of these stresses

$$\text{Thermal stress} \approx \int (\alpha_P - \alpha_S) E_P \, dT \tag{1}$$

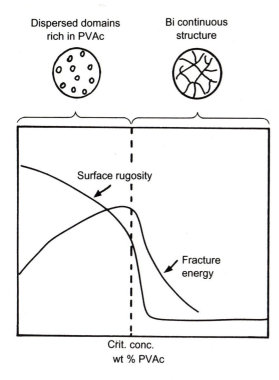

Dispersed domains rich in PVAc

Bi continuous structure

Surface rugosity

Fracture energy

Crit. conc.
wt % PVAc

Fig. 2. Typical properties of PVAc-modified UP resins as a function of the PVAc amount. At the critical concentration the morphology evolves from one containing PVAc in dispersed domains to a bicontinuous structure

where α = thermal expansion coefficient (P = polymer; S = silicon chip), E_P = Young's modulus of the polymer. The use of silica filler in the polymer formulation lowers α_P, which approaches α_S, but increases E_P. To reduce E_P through dispersion of an elastomeric phase in the epoxy resin represents a convenient alternative [13].

e) Polymer-dispersed liquid crystals. Recently nematic droplets have re-emerged as a topic of interest. Polymer-dispersed liquid crystals, PDLCs, are formed [14, 15] by incorporating initially soluble liquid crystals (20–70 wt%) in a polymer matrix that may be a crosslinked epoxy resin or a UV-cured acrylate. Nematic droplets are obtained during reaction.

Although several other examples of the use of modified thermosetting polymers may be presented, those given above point out the practical importance of generating a second-phase dispersion in a thermosetting polymer.

1.2 Polymer Network Formation

Network formation is possible if at least one of the initial monomers has a functionality higher than two. According to the specificity of chemical mechanism, the reactions of network formation may be divided into two main groups:

a) polycondensation or step polymerization,
b) chain polymerization.

For the general case where the network is formed from a monomer and a comonomer (hardener), polycondensation may be represented as

$$E_{m,n} + E_{i,j} \rightarrow E_{(m+i),(n+j)} \tag{2}$$

where $E_{m,n}$ represents an oligomer containing n molecular units of monomer and m molecular units of the comonomer (hardener). The starting monomer and comonomer are represented by $E_{0,1}$ and $E_{1,0}$, respectively.

At any conversion, p, of functional groups, a mixture of the distribution of $E_{m,n}$ species with the modifier (M) constitutes a multicomponent system. A simplified description may be made by using an average species, $\overline{E_{mn}}$, as representative of the whole population, with a size (mass) varying continuously with conversion. The simplified approach regards the system as a quasi-binary mixture of $\overline{E_{mn}}$ and M, for any conversion level.

A characteristic feature of poycondensation is the continuous change in the molecular mass distribution and the relatively high conversion of functional groups at the gel point (p_{gel} = conversion where the mass average molar mass becomes infinite).

The case of chain (co) polymerization is far more complicated [16]. An initiation step produces an active centre (free radical, anion, cation, etc, ...), that leads to a fast growth through successive monomer additions (propagation step). The propagating chain ends through transfer and/or termination steps.

The propagation step may be described by the equation

$$P_{i,j,k} + P_{1,n} \rightarrow P_{(j+n-1),k} \tag{3}$$

where i represents the number of monomer units, j is the number of functional groups and k is the number of active centres in the particular macromolecular species. $P_{1,n}$ indicates a monomer containing n functional groups in its structure.

Formation of a low concentration of polyfunctional macromolecules dissolved in the monomer is a special feature of the early stage of the chain polymerization mechanism. Because of the usually high degree of polymerization of the formed chains, the gel-point conversion is relatively low. Exceptions are cases in which a very low concentration of a polyfunctional monomer (i.e. a divinyl molecule) is copolymerized with a high concentration of a difunctional monomer (i.e. a vinyl molecule), or cases where transfer or termination rates are comparable with propagation rates (i.e. homopolymerization of epoxides initiated by tertiary amines or Lewis acids).

At any conversion, p, of functional group, a mixture of the distribution of the P_i species (containing i monomer units) with the modifier (M) constitutes a multicomponent system. In this case, a simplified description may be made by using a quasi-ternary system consisting of an average molecular species $\overline{P_i}$, the remaining monomer, P_1, and the modifier M.

An important feature in which chainwise and stepwise polymerizations differ is the tendency to cyclization. In chain polymerization the chains formed are initially infinitely diluted by the monomer so that an attack by an active site of an unreacted functional group attached to a macromolecule is much more probable through an intramolecular reaction than through an intermolecular one. Therefore, in these reactions there is a strong tendency to intramolecular cyclization. In turn, as active sites become buried in the crosslinked structure, there is a multiplication of intramolecular bonds leading to the formation of a microgel particle. As the polymerization continues, a percolation of the microgel structure is eventually produced, leading to the formation of a giant macromolecule only confined by the reactor walls. At this particular conversion, (macro) gelation takes place.

Common characteristics accompanying network formation in polycondensations and chain polymeriations are the increase in molecular masses, the broadening of the molecular mass distribution and the passing through the gel point. Another event that may take place during network formation is vitrification.

1.3 Vitrification

If the polymer network formation is performed at a constant temperature (T_{cure}), vitrification takes place at a time (conversion) when the glass transition temperature (Tg) of the partially reacted system becomes equal to T_{cure}. This requires that $T_{cure} < Tg_\infty$ (ultimate glass transition temperature of the polymer network). When the system vitrifies there is a drastic drop in both the polymerization rate and the eventual diffusional processes associated with phase separation. Therefore it is important to determine the conditions where vitrification may take place.

Usually a unique relationship may be established between Tg and conversion for a particular thermosetting polymer. This means that the network structure is entirely determined by the particular conversion level, independently of the particular thermal cycle employed to attain the particular conversion. Experimental Tg vs p relationships have been reported for several thermosetting polymers, including epoxies, polyurethanes, cyanate esters, vinyl esters, etc. In many of these systems the presence of non-idealities such as unequal reactivity of functional groups or substitution effects makes the network structure vary with the selected cure cycle. However, the experimental finding of a unique relationship between Tg and p indicates that the changes produced in the network structue are not significant enough to modify the Tg vs p relationship.

Gelation and vitrification curves are usually presented together in transformation diagrams relating temperature vs time or conversion vs temperature. Examples will be presented in the next section, including the miscibility curve for modified-thermosetting polymers.

1.4 Generation of Two-Phase Morphologies

There are two main procedures to generate a second-phase dispersion in a modified-thermosetting polymer:

a) phase separation during polymerization of an initial homogeneous solution (reaction-induced phase separation),

b) dispersion of a second phase in the starting monomers.

1.4.1 Reaction-Induced Phase Separation

Preparation of rubber-modified epoxies by this procedure was reported by McGarry et al. in the late 1960s and early 1970s [17–19]. Epoxide resins based on diglycidyl ether of bisphenol A (DGEBA) cured with diamines were toughened by the addition of low-molar-mass carboxyl-terminated poly(butadiene-acrylonitrile) copolymers (CTBN). Rubbery domains precipitated in situ during cure yielded toughened epoxy materials. The replacement of rubbers by thermoplastic polymers was developed in the 1980s [20].

Most of the modifiers exhibit an upper-critical-solution-temperature behaviour (UCST), i.e. the miscibility increases with increasing temperature. As shown in Fig. 3, a formulation containing an initial volume fraction of modifier ϕ_{Mo} is initially miscible at the reaction temperature, T_r. As the reaction proceeds, the modifier becomes less miscible, for reasons to be discussed in the next section. Phase separation begins when the miscibility gap reaches the point ϕ_{Mo}, T_r. This is defined as the cloud-point conversion (p_{cp}).

A similar situation, depicted in Fig. 4, is valid for modifiers such as poly-(ether sulfone) (PES) exhibiting a lower-critical-solution temperature behaviour [21–23]. In this case, the reaction temperature T_r is located below the initial miscibility gap.

Both types of behaviour may be described in a unique way using conversion-modifier concentration coordinates [24, 25] as shown in Fig. 5. The diagram is valid for a constant temperature. A change in temperature shifts the miscibility gap as shown in Fig. 6.

In both Figs. 5 and 6, the two-phase region has been drawn only in the pre-gel region. This is arbitrary and reflects the fact that, in systems to be described in this review, most of the phase separation takes place well before gelation, although some evidence of the continuation of the phase separation process after gelation (or even vitrification) of the thermosetting polymer has been reported [26–31]. We will return to this point when analyzing the secondary phase separation process.

Phase separation in systems involving gels is of interest in the formation of networks that are porous in the dry or swollen states [32, 33]. They are obtained by a crosslinking reaction in the presence of a diluent (modifier). Phase separation occurs either in the form of macrosyneresis (deswelling of the gel being crosslinked and formation of continuous liquid and gel phases), or microsyner-

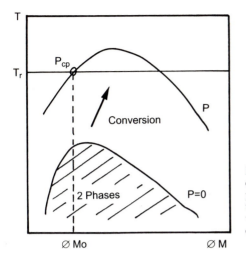

Fig. 3. Evolution of the miscibility gap with conversion (p) for a modified-thermosetting polymer showing an upper-critical-solution-temperature behaviour, UCST (ϕ_{Mo} = initial volume fraction of modifier, p_{cp} = cloud-point conversion)

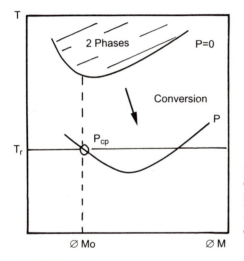

Fig. 4. Evolution of the miscibility gap with conversion (p) for a modified-thermosetting polymer showing a lower-critical-solution-temperature behaviour, LCST (ϕ_{Mo} = initial volume fraction of modifier, p_{cp} = cloud-point conversion)

esis in which the separated liquid phase forms a dispersion (droplets of liquid inside the gel) [34]. It can be caused by an increase in the degree of crosslinking (v-induced syneresis) or by changes in polymer-diluent interactions (χ-induced syneresis). In chain copolymerization of the vinyl-divinyl type, both effects are operative because the monomers act as diluents and conversion of monomers into the polymer causes not only an increase in the crosslinking density but also a change of the composition of the system undergoing polymerization [34]. A thermodynamic analysis of this kind of reaction-induced phase separation has been reported by Dušek [34–36]. The analysis has been extended to more complex situations like the equilibrium swelling of networks capable of separating into a highly swollen and a collapsed phase [37].

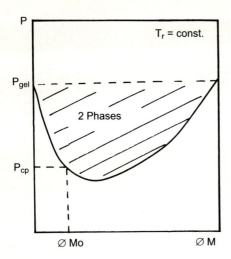

Fig. 5. Miscibility gap in conversion-modifier concentration coordinates, in the pre-gel region at a constant temperature (ϕ_{Mo} = initial volume fraction of modifier, pcp = cloud-point conversion)

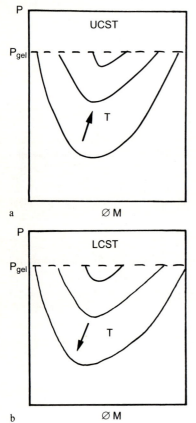

Fig. 6a, b. Shift of the miscibility gap with temperature: **a** for UCST; **b** for LCST (only the pre-gel region is shown)

The area of phase separation in systems including a gel phase is outside the scope of this review. The analysis will thus be restricted to the pregel stage of a modified thermoset.

Alternative ways of representing the reaction-induced phase separation in the case of a polycondensation (step-polymerization) process are plotted in Figs. 7 and 8 [38].

Figure 7 represents the main events taking place during the isothermal reaction of a castor oil-modified epoxy system in a time-temperature-transformation diagram (TTT) [28, 39]. It is observed that at low temperatures phase separation takes place before gelation or vitrification of the thermosetting polymer. At high temperatures, phase separation should take place after gelation. This situation could not be verified in this particular system due to the volatilization of the diamine (ethylenediamine) used as a hardener and the uncontrolled exotherms when carrying out the reaction at T > 100 °C. There is, however, experimental evidence for other systems showing no phase separation (at least in the range of sizes detected by electron microscopy), at high temperatures in the gel region [40–42].

Figure 8 shows the representation of the same events as those of Fig. 7 in a conversion-temperature-transformation diagram [43, 44]. The UCST-behaviour of the modifier is evidenced by the increase in the cloud-point conversion (p_{cp}) with temperature. Figure 9 shows a similar representation [23] for a cyanate ester system modified by a rubber (non-functionalized copolymer of

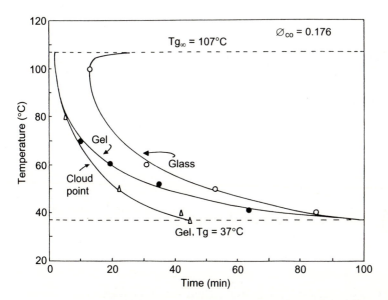

Fig. 7. TTT diagram representing times for phase separation (cloud point), gelation and vitrification for a castor-oil-modified epoxy system ($\phi_{Mo} = 0.176$) at different temperatures (Reprinted from Polymer International, **30**, R.A. Ruseckaite, L. Hu, C.C. Riccardi, R.J.J. Williams, Castor-oil-modified epoxy resins as model systems of rubber-modified thermosets. 2: Influence of cure conditions on morphologies generated, 287–295, Copyright (1993), with kind permission from the Society of Chemical Industry, London, UK)

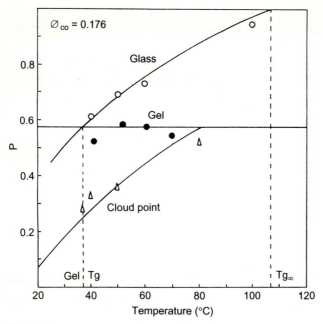

Fig. 8. Temperature-conversion transformation diagram showing phase separation, gelation and vitrification for a castor oil-modified epoxy system ($\phi_{Mo} = 0.176$) at different temperatures (Reprinted from Polymer International, **30**, R.A. Ruseckaite, L. Hu, C.C. Riccardi, R.J.J. Williams, Castor-oil-modified epoxy resins as model systems of rubber-modified thermosets. 2: Influence of cure conditions on morphologies generated, 287–295, Copyright (1993), with kind permission from the Society of Chemical Industry, London, UK)

butadiene and acrylonitrile NFBN) or a thermoplastic polymer, poly(ether sulfone), PES. While the rubber exhibits a UCST behaviour, PES shows a LCST behaviour, i.e. p_{cp} decreases as temperature increases.

All these examples (epoxy monomers cured with a diamine and reaction of cyanate-ester monomers) involve a polycondensation process. In the case of a chain polymerization, a triangular diagram is more useful to describe the miscibility curves at a constant temperature (Fig. 10). At the reaction temperature T_r, the system is passing from the one-phase region A (ϕ_{Mo}) to the two-phase region B while the miscibility gap is shifting to the monomer vertex. There are few results in the literature concerning the evolution of phase diagrams during chain polymerization of thermosets. On the other hand, the synthesis of high-impact polystyrene (HIPS) is well documented [45], and from a thermodynamic point of view, the behaviour is similar.

1.4.2 Dispersion of a Second Phase in the Starting Monomers

The second technique used in the preparation of modified-thermosetting polymers consists of introducing preformed particles in the initial formulation. This technique is also well documented in modified thermoplastics. Systems based on

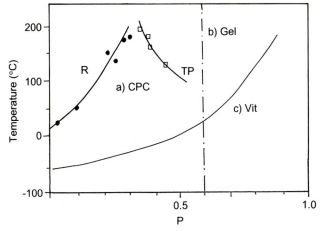

Fig. 9. Temperature-conversion transformation diagram showing phase separation, gelation and vitrification for a rubber (R) and a thermoplastic (TP)-modified cyanate ester system (15 wt% of modifier) (Reprinted from Chemistry and Technology of Cyanate Ester Resins (I. Hamerton, ed.), J.P. Pascault, J. Galy, F. Mechin, 112–150, Copyright (1994), with kind permission from Chapman & Hall, London, UK)

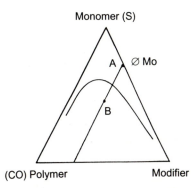

Fig. 10. Ternary phase diagram for a typical chain-polymerization (when conversion (p) increases, the trajectory follows the direction AB and the miscibility gap is shifted to the monomer vertex due to the increase in the polymer molecular mass)

dispersed acrylic elastomers [46–48] and core-shell particles, with a rubbery core embedded in a compatibilizing thermoplastic shell [49–52], have been reported. The stability of the dispersion is accomplished by adjusting the shell composition [49], or by copolymerizing the acrylic monomers with an adequately functionalized epoxide, e.g. a vinyl ester obtained by reacting the epoxy monomer with methacrylic acid [13, 47] (two sequential reactions, like in sequential IPNs).

1.4.3 Comparison Between Both Processes

Depending on applications, both procedures used to generate two-phase morphologies in modified-thermosetting polymers have their advantages and disadvantages.

Concerning the toughening of thermosetting networks, the stability and low viscosity (relative to the second procedure) of the starting homogeneous solution are the advantages of the reaction-induced phase separation. However the possibility of generating a variety of structures in the final material is the most important fact that cannot be produced by the second technique. This is, for example, formation of bicontinuous structures or generation of sandwich structures consisting of phase-separated boundary layers surrounding an homogeneous core [42]. This requires an effective control of every factor involved in the phase separation process.

Disadvantages of the reaction-induced phase separation technique are the production of undesired morphologies when using particular cure schedules, i.e. the necessity of establishing the relationships between cure cycles, morphologies generated and resulting properties. Another disadvantage is that, usually, part of the modifier remains in solution in the thermoset, even at complete conversion. If the modifier is a low-Tg (glass transition temperature) rubber, this leads to a decrease of the maximum Tg ($Tg \infty$) of the fully cured material, i.e. the modifier introduces a plasticization effect in the matrix. When the modifier is a high-Tg thermoplastic like poly(ether imide) (PEI), the residual fraction remaining dissolved in the matrix does not constitute a problem.

Regarding the dispersion of a second phase in the starting monomers, the main advantage is the initial control of the volume fraction, particle-size distribution and composition of the dispersed phase. Also important is the possibility of performing precise studies related to the influence of interfacial adhesion [49], or particle-size distribution [53], on resulting properties.

Disadvantages of the second procedure are the relatively high initial viscosity of the dispersions and the possibility of producing particle agglomeration or even macroscopic segregation during storage or processing. Besides, some of these systems give poor performance in the presence of high inorganic filler contents [54].

1.5 Scope of the Review

An efficient use of the reaction-induced phase separation process in modified-thermosetting polymers needs a deep knowledge of factors that control the morphologies generated and their relationships with the required properties. Several research groups (particularly those of the authors) have been involved in this area during the last decade. Some problems have been solved but there are still many questions to answer. At this stage, it was considered useful to summarize the present state of knowledge in this field.

Events involved in the reaction-induced phase separation process are determined by thermodynamic and kinetic factors. A thermodynamic analysis enables one to determine regions, in conversion-composition coordinates, where the system remains stable (i.e. no phase separation is produced), metastable (i.e. phase separation may take place), or unstable (i.e. phase separation does take place).

When the system evolves through the metastable region, the possibility of starting phase separation is determined by the ratio between the intrinsic rate of phase separation process (nucleation, growth, coalescence) and the reaction rate. If the ratio is high enough, a new phase will be formed with a composition that may be predicted by the thermodynamic analysis. If the ratio is low enough, no phase separation will be observed until the unstable region is attained. At this point spinodal demixing will occur.

Both thermodynamic and kinetic factors determine the morphologies generated at any point in the material. Once the influence of different variables is rationalized, the types of morphologies produced may be varied by adjusting the formulation (i.e. nature and concentration of modifier) or the cure cycle (temperature vs time).

The following sections deal with the thermodynamic basis of the phase separation process, the analysis of factors that control the morphologies generated and a brief discussion of the properties obtained as a function of the structures developed. Concluding remarks are focused on the questions that are still waiting for an answer.

2 Thermodynamic Description of the Phase Separation Process

Classic thermodynamic equilibrium considerations lead to useful descriptions of the phase equilibria in partially miscible polymer systems. According to Kleintjens and Koningsveld [55], four different levels of approximation of polymer-solvent interactions may be defined.

a) First level: solubility parameter theory that predicts whether a blend is miscible or not. There are no answers to molar mass or concentration dependence.
b) Second level: the Flory–Huggins–Staverman model (usually referred to as Flory–Huggins model, FH) that can deal with temperature and molar mass but predicts concentrations wrongly; it sets excess volumes equal to zero and thus cannot deal with pressure, expansion or contraction upon mixing.
c) Third level: FH model extended by contact statistics and equation-of-state terms (mean-field lattice-gas model). This gives a better prediction of phase concentrations and makes the influence of pressure manageable.
d) Fourth level: extended FH model taking into account nonuniform segment density and chain flexibility.

The following analysis is placed on the second level approximation, with one extra assumption, i.e. the Flory–Huggins interaction parameter χ is considered to be concentration-independent and only a function of the reciprocal of the temperature: $\chi = a + b/T$.

The system to be modelled will also be considered in two approximation levels.

First level: every constituent is considered as monodisperse. For a modi-fied-thermosetting polymer undergoing polycondensation, this approximation level leads to a quasi-binary system consisting of

$\begin{cases} \text{component P (average species of the polymer)} \\ \text{component M (average species of the modifier).} \end{cases}$

While the size (mass) of component M remains constant throughout the reaction, the size (mass) of component P increases with conversion. The selection of the average size of P and M is arbitrary. As the analysis is carried out in the pre-gel stage where every species remains finite, number-, mass- or any other average molar masses may be taken to illustrate the major trends involved in the phase separation process. Number-average molar masses will be taken in our analysis.

For a modified-thermosetting polymer undergoing chain polymerization, this approximation level leads to a quasi-ternary system consisting of

$\begin{cases} \text{monomer } P_1 \\ \text{Polymer P (average species of the polymer)} \\ \text{modifier M (average species of the modifier).} \end{cases}$

In this case, while the sizes of P_1 and M remain constant throughout the reaction, the size of P will increase in a way that depends on the particular mechanism of polymer network formation. Both the ratio of (P_1/P) and the size of P will change with the monomer conversion.

Second level: every constituent (polymer and modifier) is considered as polydisperse. In this case, while the molar mass distribution of modifier remains constant, the corresponding polymer distribution varies with the conversion level.

The FH model will be applied to a modified-thermosetting polymer under-going polycondensation. First, both constituents will be considered as mono-disperse (quasi-binary approach). Then the analysis will be performed taking polydispersity of both constituents into account. While the first analysis will enable us to discuss the most significant aspects of the phase separation process, the more refined model will provide an explanation of some experimental obser-vations like the presence of a secondary phase separation process.

2.1 Flory–Huggins Model for Monodisperse Constituents

2.1.1 Model Description

The system under analysis consists of a mixture of N_P moles of a polymer, P, undergoing polycondensation, and N_M moles of a modifier, M. The Flory–

Huggins expression for the entropy of mixing is given by [56]

$$\Delta S^M = - R(N_P \ln \phi_P + N_M \ln \phi_M) \tag{4}$$

where R is the gas constant and ϕ_i is the volume fraction of component i in the mixture.

By calling V_P and V_M the molar volumes of polymer and modifier, and $V_T = N_P V_P + N_M V_M$ the total volume of the system, the entropy of mixing per unit volume may be written as:

$$\Delta S = \frac{\Delta S^M}{V_T} = - R \left(\frac{\phi_P}{V_P} \ln \phi_P + \frac{\phi_M}{V_M} \ln \phi_M \right) \tag{5}$$

while V_M remains constant during the polycondensation of P, and V_P increases with conversion (p), following a particular law that depends on the functionality of reactants. Thus

$$V_P = V_{PO} f(p) \tag{6}$$

where V_{PO} is the initial molar volume of the monomer (or of the monomer-comonomer combination taken as a pseudocomponent).

As V_P increases with conversion, the absolute value of the entropy of mixing decreases, i.e. there is less contribution of the entropy of mixing to the free energy.

In order to express the enthalpy of mixing we define a unit cell with a molar volume, V_r (reference volume). This may be selected as the molar volume of the starting monomer(s), V_{PO}, or as the molar volume of the repetitive unit of the modifier (when M is a polymer), etc.

The number of moles of unit cells in the system is V_T/V_r. The probability of having a contact between P and M is given by [56]

$$\text{contact } P - M = Z_c \phi_P \phi_M (V_T/V_r) \tag{7}$$

where Z_c is the coordination number of a unit cell.

The exchange energy is defined as [56]

$$\varepsilon = \tfrac{1}{2} \varepsilon_{PP} + \varepsilon \tfrac{1}{2} \varepsilon_{MM} - \varepsilon_{PM} \tag{8}$$

where ε_{ij} is the energy of a contact between components i and j.

Then the enthalpy of mixing may be expressed as

$$\Delta H^M = Z_c \varepsilon (V_T/V_r) \phi_P \phi_M \tag{9}$$

or as an intensive property per unit volume of the system

$$\Delta H = (Z_c \varepsilon / V_r) \phi_P \phi_M. \tag{10}$$

The Flory–Huggins interaction parameter is defined as [56]

$$\chi = Z_c \varepsilon / RT. \tag{11}$$

Taking into account Eqs. (5), (10) and (11), the Gibbs free energy per unit volume of system is given by

$$\Delta G = \Delta H - T\Delta S = RT \left\{ \frac{\phi_P}{V_P} \ln \phi_P + \frac{\phi_M}{V_M \ln \phi_M} + \frac{\chi}{V_M \ln \phi_M} + \frac{\chi}{V_r} \phi_P \phi_M \right\}.$$

(12)

In the frame of this approach, χ is taken as an exclusive function of temperature:

$$\chi = a + b/T.$$

(13)

An equivalent definition of an interaction parameter with units of energy per unit volume is

$$\Lambda = \chi RT/V_r.$$

(14)

In the first level of the thermodynamic description, Λ may be related to the difference between the solubility parameters of both components, i.e.

$$\Lambda = (\delta_P - \delta_M)^2.$$

(15)

In more refined models, the interaction parameter is defined as a function of both temperature and composition. Usually it is expressed as a g (T, ϕ_M) function that replaces χ in Eq. (12). There is also an effect of molar mass but it is typically small [57]. For example, Koningsveld and Kleintjens [58] proposed the expression

$$g = g_0 + g_1/(1 - \gamma \phi_M)$$

(16)

where g_0 and γ are constants and g_1 is taken as a function of temperature, e.g. $g_1 = g_{10} + g_{11}/T$.

An interaction parameter depending on both temperature and composition may also be defined in terms of the chemical potential of one of the components (P in our case) [57, 59]. The expression

$$\chi = [1 + b_1 \phi_M + b_2 \phi_M^2] \, [d_0 + d_1/T + d_2 \ln(T)]$$

(17)

where the coefficients b_i and d_i are adjustable parameters may be used to represent the five most common types of experimentally observed phase diagrams: LCST, UCST, combined LCST/UCST with the former occurring at higher temperature than the latter, hourglass, and closed loop [60, 61].

The interaction parameters g (T, ϕ_M), defined in the free energy function, and χ (T, ϕ_M), defined in the expression of the chemical potential of one of the components (i.e. P), are related by [57, 60]:

$$\int_{\phi_M}^{1} \chi(T, \phi) d\phi = (1 - \phi_M) g(T, \phi_M).$$

(18)

Returning to Eq. (12) (with χ given by Eq. 13), the following definitions may be introduced:

$$Z_P = V_P/V_r = Z_p(p)$$

(19)

$$Z_M = V_M/V_r.$$

(20)

This leads to

$$\Delta G = (RT/V_r) \left\{ \frac{\phi_P}{Z_P(p)} \ln \phi_P + \frac{\phi_M}{Z_M} \ln \phi_M + \chi(T) \phi_P \phi_M \right\}. \tag{21}$$

Usually $Z_M > Z_P(o)$, meaning that the contribution of the monomer (or monomer-comonomer combination) to the entropy of mixing, is much more important in absolute value than the modifier contribution. Therefore, the increase of Z_P with conversion process is feasible. For this reason it is usually stated that phase separation is the result of the decrease in the entropic contribution to the free energy of mixing during polymerization [24].

Alternatively, as some chemical groups are continuously transformed into different chemical groups during polymerization, the interaction parameter between both components may vary throughout the reaction, i.e. ε in Eq. (11) may be a function of conversion. Thus, χ may increase (favouring demixing), remain constant or decrease (favouring mixing) during polymerization. The dependence of χ on conversion can be modelled assuming that each segment can have different kinds of interacting sites of total number Z_c. This requires the use of a quasi-chemical equilibrium approach and Tompa's modification for polymer molecules with different contact points [62, 63].

2.1.2 Binodal Curve

The binodal curve is the boundary between thermodynamically stable and metastable solutions. The term binodal is used in truly binary systems while in actual polydisperse systems the correct denomination is cloud-point curve (CPC). Thus the experimental determination of this boundary always leads to a CPC.

When using a quasi-binary model as a representation of the system, a binodal curve may be predicted (if it is desired to fit the CPC, the polydispersity of both components must be taken into account).

For the system to consist of two phases in equilibrium, the condition described by the binodal, the chemical potential of a component, must be the same in both phases. The chemical potential (partial molar free energy) may be calculated in terms of the extensive value of the free energy of mixing, $\Delta G^M = \Delta H^M - T\Delta S^M$, arising from Eqs. (4) and (9). It is defined as

$$\Delta \mu_i = \frac{\partial \Delta G^M}{\partial N_i} \bigg|_{N_j, T, \text{pressure}}. \tag{22}$$

For the solution of a thermosetting polymer (P) with a modifier (M), chemical potentials are given by

$$\Delta \mu_P = [RT Z_P(p)] \left\{ \frac{1}{Z_P(p)} \ln \phi_P + \left[\frac{1}{Z_P(p)} - \frac{1}{Z_M} \right] \phi_M + \chi \phi_M^2 \right\} \tag{23}$$

$$\Delta\mu_M = [RTZ_M]\left\{\frac{1}{Z_M}\ln\phi_M + \left[\frac{1}{Z_M} - \frac{1}{Z_P(p)}\right]\phi_P + \chi\phi_P^2\right\}. \tag{24}$$

Then the equilibrium condition (binodal curve) arises from

$$\Delta\mu_P^\alpha = \Delta\mu_P^\beta \tag{25}$$

$$\Delta\mu_M^\alpha = \Delta\mu_M^\beta \tag{26}$$

where α and β represent the two phases in equilibrium.

Fixing T (and consequently χ), and $\phi_P^\alpha(\phi_M^\alpha = 1 - \phi_P^\alpha)$ Eqs. (25) and (26) may be solved to obtain the conversion (p) and the composition of the β-phase, $\phi_P^\beta(\phi_M^\beta = 1 - \phi_P^\beta)$. In a representation of conversion vs composition at constant temperature, compositions at equilibrium, i.e. ϕ_M^α and ϕ_M^β, will be joined by horizontal tie lines at conversion.

2.1.3 Spinodal Curve

The spinodal curve constitutes the boundary between thermodynamically metastable and unstable regions. At this critical boundary, the following condition is fulfilled [56]

$$\left.\frac{\partial\Delta\mu_i}{\partial\phi_j}\right|_{T,\text{pressure}} = 0. \tag{27}$$

For the thermosetting polymer (P)-modifier (M) system, the spinodal curve satisfying Eq. (27), is given by

$$\frac{\phi_M}{Z_P(p)} + \frac{(1 - \phi_M)}{Z_M} = 2\chi\phi_M(1 - \phi_M). \tag{28}$$

The quadratic equation can be solved for p as a function of ϕ_M, i.e. the spinodal curve in conversion vs composition coordinates.

Experimental measurements of the spinodal curve require bridging of the metastable region of the miscibility gap, i.e. traversing this region without phase separation. A nice technique developed for this purpose is the Pulse Induced Critical Scattering (PICS) [64].

2.1.4 Critical Point

The point where the binodal and spinodal are cotangent is called the critical point. This occurs when the first and second derivatives of the chemical potential are both zero. This gives

$$\phi_{M\text{crit}} = \frac{1}{1 + [Z_M/Z_P(p_{\text{crit}})]^{1/2}} \tag{29}$$

and

$$\chi = \frac{1}{2} \left\{ \frac{1}{[Z_P(p_{crit})]^{1/2}} + \frac{1}{Z_M^{1/2}} \right\}^2.$$ (30)

The critical point in conversion vs composition coordinates (ϕ_{Mcrit}, p_{crit}), is obtained from Eqs. (29) and (30).

2.1.5 Illustration

In order to obtain a phase diagram in conversion-composition coordinates, a particular system consisting of a diepoxide based on diglycidylether of bisphenol A (DGEBA), a stoichiometric amount of a diamine (4,4'-diamino-3,3'-dimethyldicyclohexyl-methane, 3DCM), as a hardener, and a rubber based on a statistical copolymer of butadiene and acrylonitrile, will be considered [65]. The increase in the number-average molar mass for the polycondensation of a stoichiometric diepoxy-diamine mixture, is given by [65]

$$\overline{M_n/M_{n0}} = V_P/V_{P0} = 1/(1 - 4p/3).$$ (31)

By defining $V_r = V_{P0}$ in Eq. (19), $Z_P = 1/(1 - 4p/3)$.
Besides $Z_M = 13.73$ and χ (at 348 K) = 0.63 $\pm$ 0.03 (as obtained from the fitting of experimental and predicted cloud-point conversions [65]).

The binodal and spinodal curves can be calculated from Eqs. (25), (26) and (28). Results are plotted in Fig. 11 in the conversion range located in the pregel stage ($p_{gel} = 0.60$). As the value of $\overline{M_n}$ has no physical sense beyond this limit, calculations should not be extrapolated.

Figure 11, although based on an extremely simple model, provides a qualitative explanation of the main facts involved in the phase separation process. Outside the binodal, the system remains as a stable homogeneous solution. The area between the binodal and spinodal curves is the metastable region. If kept there, phase separation will proceed kinetically by a nucleation-growth (NG) mechanism provided that the phase separation rate is sufficiently higher than the polymerization rate. Inside the spinodal the system is unstable and spontaneously decomposes by the so-called spinodal demixing process (SD). The process is unstable, requires no activation energy, and is characterized by a negative value of the diffusion coefficient [56].

By replacing $Z_P(p)$ in Eqs. (29) and (30), critical coordinates are given by

$$\phi_{Mcrit} = 1/\{1 + [Z_M(1 - 4p_{crit}/3)]^{1/2}\}$$ (32)

$$p_{crit} = (3/4)\{1 - [(2\chi)^{1/2} - Z_M^{-1/2}]^2\}.$$ (33)

Replacing the particular values of χ and Z_M characterizing the selected rubber-modified epoxide system gives $\phi_{Mcrit} = 0.24$ (or 0.212 as wt%) and $p_{crit} = 0.205$. Therefore formulations containing an initial amount of modifier $\phi_{Mo} < \phi_{Mcrit}$ will lead to a separated phase richer in the modifier (compositions

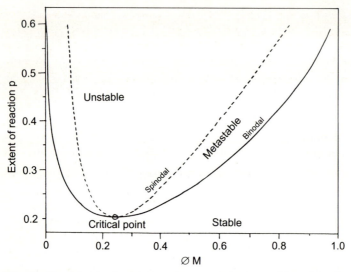

Fig. 11. Binodal and spinodal curves in a conversion vs modifier volume fraction phase diagram. The critical point and the location of stable, metastable and unstable regions are shown

of phases in equilibrium are obtained from horizontal tie lines in Fig. 11), whereas formulations with $\phi_{Mo} > \phi_{Mcrit}$ will lead to a phase inversion, i.e. the separated phase will be richer in the thermosetting polymer producing a reinforced rubber instead of a rubber-modified thermoset.

At this point we are assuming a very fast phase separation rate such that the system attains the equilibrium condition for every conversion. At the conversion where the binodal curve is first attained (p_{cp} = cloud-point conversion), a differential amount of a new phase, with a composition located on the other branch of the binodal, appears in the system. In this sense, it is correct to assess that the new phase will be the dispersed one. However, as polycondensation of the thermosetting polymer proceeds, a macroscopic amount of the initially dispersed phase will be present in the system. At this stage, the distinction between continuous and dispersed phases will depend on other factors such as interfacial energies, viscosities, volume ratio of phases, etc.

In order to obtain a β-phase rich in modifier dispersed in the thermosetting polymer, the convenient condition (albeit not sufficient) is to select an initial modifier concentration, $\phi_{Mo} < \phi_{Mcrit}$. Usually this condition is fulfilled when using 15 wt% for a modifier with a number average molar mass, $\overline{M_n} < 5000 \text{ g mol}^{-1}$ (commercial rubbers used to modify thermosetting polymers meet this condition). However, when thermoplastic polymers are used as modifiers, there is a significant increase in the range of molar masses that are normally employed [20]. From Eqs. (32) and (33) the critical modifier concentration can be also written as

$$\phi_{Mcrit} = 1/(2\chi Z_M)^{1/2}. \tag{34}$$

This means that ϕ_{Mcrit} decreases with the square root of the modifier molar mass. Now, a 15 wt% modifier may be located on the other side of the critical point leading to a phase inversion (at least in the initial stages of the phase separation process). The relatively high increase of fracture energy with the molar mass of the thermoplastic modifier, experimentally observed [20], may be ascribed to the phase inversion process.

Another important point arising from Fig. 11 is the fact that, during the initial stages of phase separation (i.e. at conversions well before gelation), the dispersed β-phase contains a significant amount of thermosetting polymer. This is a direct consequence of thermodynamics and does not depend on a possible reaction between end groups of the modifier and the thermosetting polymer, as is often stated.

As χ is assumed constant in the calculation of binodal and spinodal curves, the only factor promoting phase separation is the increase of V_P with conversion. For example, a cloud-point conversion $p_{CP} = 0.25$ means that phase separation begins when $(V_P/V_{P0}) = 1.5$, i.e. a 50% increase in the molar volume of the thermosetting polymer is enough to reach the binodal curve. At this regard, it has been shown [66] that the location of the miscibility gap in non-reactive binary mixtures of a diepoxide, based on DGEBA and a rubber based on a carboxyl-terminated butadiene-acrylonitrile copolymer (CTBN), is highly dependent on the molar mass of the diepoxide. Cloud-point curves (CPC) in temperature vs composition coordinates were obtained for diepoxides of different number-average molar masses, i.e. 349, 383, 479 and 550 g mol^{-1}. The maximum of the CPC (precipitation threshold temperature) was located at 51, 65, 116 and 135 °C, respectively [66]. These results constitute clear evidence of the significant influence of the molar volume of the thermosetting polymer on the location of the miscibility gap. The simplified model under consideration predicts the observed trend; a quantitative fitting of the CPC would require one to take polydispersity of constituents into account.

2.1.6 Factors that Control Miscibility

In the development of modified thermosetting polymers undergoing a reaction-induced phase separation process, it is important to select a modifier such that phase separation precedes gelation for reasons to be discussed later. Once the thermosetting polymer has been defined, the selection of a suitable modifier has to take into account such factors as type and composition (i.e. for a copolymer the comonomers, their relative fraction and the comonomers distribution, statistical, block, graft, etc.), nature of end-groups, and molar mass distribution. All these factors have a strong effect on the location of the miscibility gap (through the factors χ and Z_M in the simple thermodynamic model under consideration).

Regarding the influence of the nature of the modifier on its miscibility with the starting monomers, a vast literature is available in the field of rubber-modified epoxies [66–71]. Most of these studies have been carried out in systems devoid of hardener, i.e. in non-reactive rubber-diepoxide systems. In some cases, results for systems consisting of a rubber dissolved in an off-stoichiometric epoxy-amine system, (i.e. using a stoichiometric imbalance so that gelation is avoided at full conversion) have also been reported. These experimental results give consistent trends which make it possible to predict the miscibility in stoichiometric systems, i.e. to order the modifiers in a series varying from less miscible to more miscible.

For example, it has been shown that by increasing the polarity of the rubber through incorporation of polar comonomers in the chain, the miscibility with a DGEBA-based epoxy resin increases. Thus, an increase by 1% in acrylonitrile (AN) content in a statistical copolymer with butadiene leads to a decrease of about 10 K in the precipitation threshold temperature (maximum of the experimental CPC) [71]. The nature of the end-group of these low molar mass copolymers also has a strong influence on the miscibility with a DGEBA-based epoxy. It was found [71] that the miscibility of DGEBA with rubbers end-capped by different functional groups decreases in the series

$$ATBN > ETBN > CTBN > NFBN.$$

The abbreviations mean amino-, epoxy-, carboxyl- and non-functional end-capped polybutadiene-acrylonitrile copolymers, respectively, of the same molar mass and 18% AN content. The polarity of rubbers changes in the same order except for CTBN. Lower values of CTBN miscibility in comparison with ATBN and ETBN can possibly be caused by the great self-association effect of carboxyl groups through hydrogen bonding with the formation of high molar mass chains.

As predicted by Eq. (33), an increase in the modifier molar mass, i.e. increase in Z_M, shifts the phase separation region to lower conversions (there is a decrease in $p_{crit.}$). Figure 12 shows binodal curves calculated for an epoxy-diamine system, modified by three butadiene-acrylonitrile copolymers of different number-average molar mass [72]. These curves were calculated in a similar way as the curves shown in Fig. 11, i.e. both components were regarded as monodisperse with a molar volume determined by the number-average molar mass (constant for the butadiene-acrylonitrile copolymer and variable with conversion for the epoxy-diamine system). The interaction parameter for each particular system was fitted by making the predicted binodal curves to pass through the experimental points (ϕ_{Mo}, p_{cp}).

Therefore, once the type of modifier is selected, e.g. a statistical copolymer of acrylonitrile and butadiene, its miscibility with the monomers may be varied significantly by changing the nature of end-groups, the fraction of both comonomers in the copolymer, and the molar mass distribution. The same trends have been observed with rubbers based on n-butyl acrylate [73] or dimethylsiloxane [74]. The introduction of acrylonitrile units, polycaprolactone blocks [75], or phenyl groups in the modifiers increases the polarity and the initial miscibility in polar monomers like epoxy resins or bismaleimides [76].

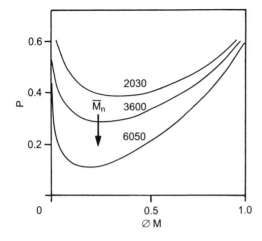

Fig. 12. Binodal curves calculated for an epoxy-diamine system (DGEBA-3DCM), modified by 15 wt% of three butadiene-acrylonitrile copolymers of different number-average molar mass (DGEBA = diglycidyl ether of bisphenol A; 3 DCM = 4,4'-diamino-3,3'-dimethyl-dicyclohexyl methane)

2.1.7 Application to Particular Systems

The simple thermodynamic model derived in Sect. 2.1 has been useful to get a qualitative insight into the phase separation process. When one intends to apply it to an actual system, the significant influence of polydispersity is clearly evidenced. For example, Fig. 13 shows the experimental cloud-point curve for a DGEBA-CTBN binary mixture, together with binodal and spinodal curves calculated by assuming monodisperse components [66] (curves are arbitrarily fitted to the critical point). The shape of the CPC and precipitation threshold temperature (maximum of the CPC) appearing at low modifier concentrations are a clear manifestation of the rubber polydispersity [77].

In order to analyze the applicability of the thermodynamic model to an actual system, epoxy-amine formulations modified by castor oil (CO) were studied [78]. Castor oil is basically a monodisperse modifier with $\overline{M_n} = 928$ g mol^{-1} (a slight variation about this average may be produced by variations in the amount of ricinoleic acid (88% of fatty acids), linoleic acid, oleic acid and other fatty acids present in the system). Its molar volume is $V_M = 966.7$ cm^3 mol^{-1}.

It was found that a CO-modified DGEBA ($\overline{M_n} = 368$ g mol^{-1}) cured with a stoichiometric amount of ethylenediamine (EDA) segregated in the course of polymerization. The CPC of the initial system is shown in Fig. 14.

As DGEBA and EDA are used in stoichiometric proportions, a pseudo-component with a number average molar mass equal to $(M_{EDA} + 2\,M_{DGEBA})/3$ and a molar volume $V_P = 226.7$ cm^3 mol^{-1}, may be defined. From Eq. (29), the critical composition of the initial system before reaction is $\phi_{Mcrit} = 0.326$, which does not agree with the maximum of the CPC (Fig. 14). A correct fitting of the experimental CPC would require an interaction parameter depending on composition or three interaction parameters, e.g. one for each couple of components.

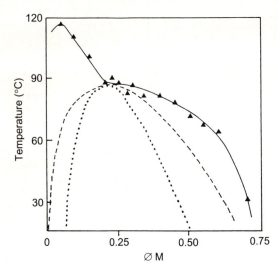

Fig. 13. Experimental CPC (▲) and calculated binodal (*broken curve*) and spinodal (*dotted curve*) curves for a binary system composed of a DGEBA-based epoxy monomer ($\overline{M}_n = 479\ \text{g mol}^{-1}$) and a carboxyl-terminated butadiene-acrylonitrile rubber (CTBN) (Reprinted from Polymer, **30**, D. Verchère, H. Sautereau, J.P. Pascault, S.M. Moschiar, C.C. Riccardi, R.J.J. Williams, Miscibility of epoxy monomers with carboxyl-terminated butadiene-acrylonitrile random copolymers, 107–115, Copyright (1989), with kind permission from Butterworth–Heinemann journals, Elsevier Science Ltd, The Boulevard, Langford Lane, Kidlington OX5 1GB, UK)

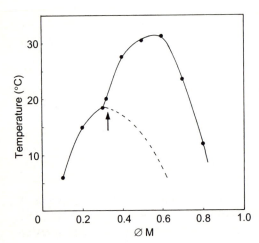

Fig. 14. Cloud-point temperature vs volume fraction of CO in a mixture with a stoichiometric DGEBA-EDA (ethylene diamine) system before reaction. *The arrow* indicates the critical-point composition predicted considering DGEBA-EDA as one pseudocomponent. *The dotted line* represents the branch rich in CO predicted using the pseudocomponent approximation (Reprinted from Polymer International, **30**, R.A. Ruseckaite, R.J.J. Williams, Castor-oil-modified epoxy resins as model systems of rubber-modified thermosets. 1: Thermodynamic analysis of the phase separation, 11–16, Copyright (1993), with kind permission from the Society of Chemical Industry, London, UK)

Even if the pseudo-component approach does not fit the whole CPC it is possible to use this model to adjust the branch poor in CO, i.e. that of practical interest, leading to a dispersed phase rich in CO and generating a shadow curve (dotted line in Fig. 14). The interaction parameter resulting from the fitting was [78]

$$\chi = -\,2.037 + 913.7/T. \tag{35}$$

Now, keeping the value of χ constant during the epoxy-amine reaction, the cloud-point conversion for a particular cure temperature may be predicted by

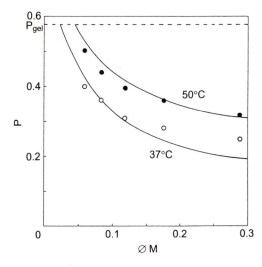

Fig. 15. Cloud-point conversion as a function of the volume fraction of CO in a DGEBA-EDA system for two different reaction temperatures. *Full lines* are theoretical predictions from the Flory–Huggins equation using $\chi(T)$ from Eq. (35) (Reprinted from Polymer International, **30**, R.A. Ruseckaite, R.J.J. Williams, Castor-oil-modified epoxy resins as model systems of rubber-modified thermosets. 1: Thermodynamic analysis of the phase separation, 11–16, Copyright (1993), with kind permission from the Society of Chemical Industry, London, UK)

solving Eqs. (25) and (26) with $Z_P(p)$ arising from Eq. (31) and $Z_M = 4.26$. Figure 15 shows a reasonable fitting between experimental values of p_{cp}, obtained at two different temperatures, and the theoretical prediction without using any adjustable parameter [78].

The experimental evidence supports the fact that phase separation is produced by the decrease in the entropic contribution to the free energy of mixing during the epoxy-amine polycondensation.

2.1.8 Limitations of the Model

As previously discussed, the model considering monodisperse constituents is the simplest thermodynamic approach to get a qualitative insight into the reaction-induced phase separation. It can explain the appearance of metastable and unstable regions in the pre-gel stage due to the increase in the average size of the growing thermosetting polymer. The phase-inversion phenomenon is also adequately predicted as well as the fact that the dispersed phase contains a significant fraction of thermosetting polymer.

However, a major limitation of this model is the impossibility of fitting cloud-point curves for polydisperse systems. Moreover, it cannot deal with the fractionation effect accompanying phase separation, i.e. the dispersed phase will be enriched in the highest molar-mass fractions of modifier but in the lowest molar-mass fractions of the growing thermosetting polymer. This may produce variations in stoichiometry and conversion between both phases. These phenomena can be conveniently treated taking polydispersity of constituents into account.

2.2 Flory–Huggins Model for Polydisperse Constituents

2.2.1 Model Description

The Gibbs free energy per unit volume for the mixture of a polydisperse modifier (M) with a polydisperse thermosetting polymer (P) is given by

$$\Delta G = RT \{ (1/V_{Ps}) \sum (\phi_y/y) \ln \phi_y + (1/V_{Ms}) \sum (\phi_x/x) \ln \phi_x + (\chi/V_r) \phi_P \phi_M \}$$

(36)

where V_{Ps} = molar volume of the unit segment of the y-mer of thermosetting polymer, V_{Ms} = molar volume of the unit segment of the x-mer of modifier, $\phi_P = \Sigma \phi_y$ and $\phi_M = \Sigma \phi_x$.

In order to solve Eq. (36) it is necessary to know the molar-mass distributions of modifier, $\phi_x(x)$, and thermosetting polymer, $\phi_y(y)$. The latter needs a statistical description of the network build-up.

Methods for calculating cloud-point and spinodal curves from Eq. (36) have been discussed in the literature [57, 61, 79–88]. The interaction parameter, χ, may be empirically expressed as a function of temperature, polymer volume fraction and number-average degree of polymerization [87].

In our analysis, χ will be taken as an exclusive function of temperature through Eq. (13), although it will be allowed to vary with conversion. The particular solution procedure developed by Kamide and co-workers [87] will be used for the calculation of cloud-point and spinodal curves.

A particular example is analyzed in the following section.

2.2.2 Application to Rubber-Modified Cyanate Esters

The cyclotrimerization of a cyanate ester (4,4' dicyanato-1,1' diphenylethane, CE) (Fig. 16), in the presence of a butadiene-acrylonitrile random copolymer terminated by non-functional groups (NFBN), has been analyzed [89].

The mass fraction distribution of rubber, $\omega(x)$, was described by the Schulz–Zimm (SZ) equation [90]:

$$\omega(x) = [U^{h+1}/\Gamma(h+1)] x^h \exp(-Ux)$$

(37)

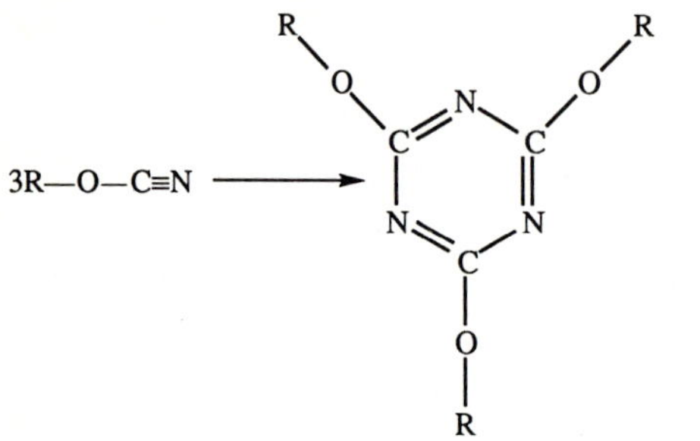

Fig. 16. Cyclotrimerization of a cyanate ester

where x is the polymerization degree taken as a continuous variable, $h = [(\overline{x_w}/\overline{x_n}) - 1]^{-1}$, $U = h/\overline{x_n}$ and Γ is the gamma function. For the selected NFBN rubber, $h = 1.087$ and $U = 0.0212$.

As the butadiene (BD) and acrylonitrile (AN) comonomers have close values of molar masses (BD $= 54$ g mol^{-1}, AN $= 53$ g mol^{-1}), one of them may be taken to define a single $V_{Ms} = 56.96$ cm^3 mol^{-1}. Also, the reference volume was taken as $V_r = V_{Ms}$.

Figure 17 [89] shows the fitting of the experimental CPC obtained by solving Eq. (36) for the initial system (the CE monomer is a solvent of the polydisperse rubber). The $\chi(T)$ function arising from the fitting was

$$\chi = -0.0372 + 68.692/T. \tag{38}$$

Now the precipitation threshold temperature is correctly predicted.

To analyse the phase separation process during reaction it is necessary to describe the distribution of oligomer species in the pre-gel stage. Assuming an ideal polycyclotrimerization [91–94], the distribution in the pre-gel stage may be described by the equation [95, 96]

$$\omega_y = \{(2y + 2)!/[(y + 2)!(y + 1)!]\}\, p^y(1 - p)^{y + 2} \tag{39}$$

where y is the number of triazine rings per molecule (the species with $y = 0$ is the unreacted monomer), ω_y represents the mass fraction of the y-mer and p is the conversion of cyanate groups. In Eq. (36), (ϕ_y/y) must be replaced by $[\phi_y/(2y + 1)]$ to take into account the fact that we are dealing with a cyclo-trimerization. CPC and spinodal curves were obtained from Eq. (36) using the method developed by Shirataki et al. [97].

Experimental values of cloud-point conversions were obtained by advancing the reaction at $T > 150\,°C$ (without a catalyst, typically $T = 180\,°C$), and slowly

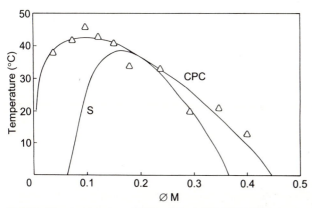

Fig. 17. Cloud-point curve (CPC) of the initial CE-NFBN system. *Full lines* represent CPC and spinodal curve (S) fitted using the Flory–Huggins equation (CE = cyanate ester, NFBN = non-functionalized polybutadiene-acrylonitrile copolymer) (Reprinted from Polymer, **36**, J. Borrajo, C.C. Riccardi, R.J.J. Williams, Z.Q. Cao, J.P. Pascault, Rubber-modified cyanate esters: thermodynamic analysis of phase separation, 3541–3547, Copyright (1995), with kind permission from Butterworth–Heinemann journals, Elsevier Science Ltd, The Boulevard, Langford Lane, Kidlington OX5 1GB, UK)

cooling in a light transmission device [89]. The cloud-point temperature was thus obtained in situ while the corresponding cloud-point conversion, p_{cp}, was determined from the residual reaction heat measured by differential scanning calorimetry.

The fitting of experimental values with the theoretical solution derived from Eq. (36) required that the interaction parameter decreases with conversion according to the relationship [89]

$$\chi = -0.372 + b(p)/T. \tag{40}$$

The decrease of the parameter b with conversion is represented in Fig. 18 (experimental points were obtained for a range of rubber mass fractions comprised between 3 and 20%). The decrease of χ with conversion may be ascribed to the transformation of polar cyanate groups into a less polar ring structure which becomes more miscible with the NFBN rubber.

Another set of experimental results was obtained by operating at a constant temperature (T = 180 °C) in the light transmission device. Cloud points are representative of the isothermal reaction-induced phase separation process. Experimental points could be fitted by solving Eq. (36), using the $\chi(p)$ relationship arising from Eq. (40). Figure 19 shows a reasonable prediction of the experimental CPC using the $\chi(p)$ functionality. Also shown is the CPC, calculated by assuming that χ keeps the initial value given by Eq. (38). It is clear that the polycyclotrimerization reaction affects the rubber miscibility in two different ways: a) there is a decrease in the entropic contribution to the free energy of mixing, a factor that decreases miscibility; b) there is a decrease in the value of the interaction parameter derived from the change in chemical structure, a factor that favours miscibility. Demixing takes place because of the

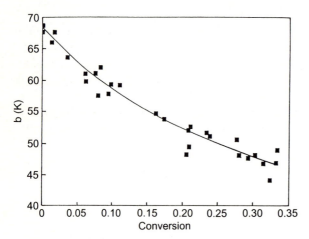

Fig. 18. Variation of the parameter b(K) with conversion arising from the fitting of experimental cloud-point conversions obtained for NFBN-CE formulations containing rubber mass fractions between 0.03 and 0.20 (Reprinted from Polymer, **36**, J. Borrajo, C.C. Riccardi, R.J.J. Williams, Z.Q. Cao, J.P. Pascault, Rubber-modified cyanate esters: thermodynamic analysis of phase separation, 3541–3547, Copyright (1995), with kind permission from Butterworth–Heinemann journals, Elsevier Science Ltd, The Boulevard, Langford Lane, Kidlington OX5 1GB, UK)

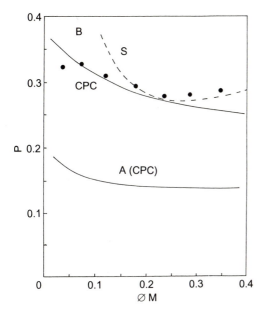

Fig. 19. Experimental cloud points and CPC and spinodal (S) curves calculated with the Flory–Huggins equation, assuming that the interaction parameter decreases with conversion according to Eq. (40). Curve A represents the CPC calculated using the initial value of the interaction parameter, assumed constant (Reprinted from Polymer, **36**, J. Borrajo, C.C. Riccardi, R.J.J. Williams, Z.Q. Cao, J.P. Pascault, Rubber-modified cyanate esters: thermodynamic analysis of phase separation, 3541–3547, Copyright (1995), with kind permission from Butterworth–Heinemann Journals, Elsevier Science Ltd, The Boulevard, Langford Lane, Kidlington OX5 1GB, UK)

prevailing effect of the first factor. The second factor shifts the CPC to higher conversions.

Therefore it may be stated that the reaction-induced phase separation process in a modified thermosetting polymer is produced by the increase in the average polymer size superimposed to a possible variation of the interaction parameter with conversion. This second effect may favour mixing or demixing, depending on whether χ decreases or increases.

An interesting fact was observed (Fig. 19) in that, for solutions containing less than 0.12 volume fraction of rubber, phase separation proceeds along the calculated CPC. However, for compositions close to or to the right side of the critical point, phase separation is better represented by the spinodal curve. It may be inferred that the nucleation-growth (NG) mechanism is operative at low rubber volume fractions (i.e. for the usual formulations employed to generate a rubber-modified thermoset), while spinodal demixing (SD) takes place near the critical point or in the phase-inversion region.

2.2.3 Phase Separation Along Reaction: Coexistence Curves

In order to make a thermodynamic analysis of the phase separation process in the metastable region (i.e. after reaching the CPC), the castor oil-modified DGEBA-EDA system will be considered again. The resulting trends may be generalized for any modified thermosetting polymer.

The reaction between diglycidylether of bisphenol A (DGEBA) and ethylenediamine (EDA) may be considered as an ideal ($A_4 + B_2$) polycondensation [38, 98, 99]. Therefore expressions are available to describe the distribution of species generated along the reaction [90].

However, as in this section the evolution of phase separation in the metastable region will be analyzed, conditions will be such that a fractionation of the oligomer species between the continuous and dispersed phases will be produced. Therefore the available statistical distributions are of no value and it is necessary to calculate the evolution of the oligomer distributions in both phases using the kinetic equations describing the polycondensation.

The molar concentration of species generated during polymerization will be denoted as $E_{m,n}$, m and n being, respectively, the number of diamine and diepoxide molecules present in a particular species $(m - 1 \leq n \leq 3m + 1)$. Figure 20 shows both monomers, $E_{0,1}$ (DGEBA), $E_{1,0}$ (EDA) and one of the possible isomers of $E_{3,5}$ ($E_{m,n}$ includes all the particular isomers that may be generated with the assumption of no intramolecular reaction). The system of kinetic equations necessary to obtain the $E_{m,n}$ distribution in both phases has been reported elsewhere [100, 101]. Once the concentration, $E_{m,n}$, and volume fraction, $\phi_{m,n}$, of oligomer species is known, the Gibbs free energy per unit volume can be stated:

$$\Delta G = (RT/V_{1,0})\left\{\sum_m \sum_n (\phi_{m,n}/Z_{m,n})\ln \phi_{m,n} + (\phi_M/Z_M)\ln \phi_M + \chi \phi_P \phi_M\right\}$$

(41)

where $V_{1,0}$ (molar volume of EDA), is taken as the reference volume and

$$\phi_P = \sum_m \sum_n \phi_{m,n}$$

(42)

$$Z_M = V_M/V_{1,0}$$

(43)

$$Z_{m,n} = V_{m,n}/V_{1,0}$$

(44)

$V_{m,n}$ being the molar volume of $E_{m,n}$.

In order to calculate CPC and spinodal curves from Eq. (41) it is necessary to define the value of the interaction parameter. A simulation was performed for an

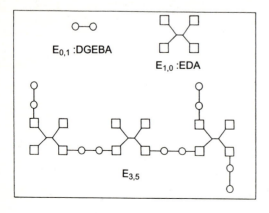

Fig. 20. Monomers and one of the oligomer species generated during the DGEBA-EDA step copolymerization

isothermal reaction, assuming a constant χ value independent of conversion. Details of the calculations have been reported elsewhere [101]. Only the most significant results will be discussed in what follows.

Figure 21 shows the composition and conversion of both phases (α = continuous, β = dispersed), when complete equilibrium is obtained for any conversion level. This is a limiting situation where the mass transfer rate between phases, including the generation of a new phase, is much higher than the polymerization rate. It is observed that conversion in the β-phase is significantly lower than the overall conversion of epoxy (and amine) groups. This is due to the preferential segregation of monomers and low-molar mass species (also with lower functionalities) to the β-phase. This has an important consequence related to the fact that the step-polymerization in the β-phase may continue for a long time (conversion range) after gelation (or even vitrification) of the α-phase. The generation of a dispersed β-phase in an α-matrix is usually referred to as the development of the primary morphology.

Figure 22 shows the increase in the volume fraction of dispersed β phase, after the cloud point is reached, as a function of the overall conversion of epoxide groups in the system. Most of the primary morphology is generated in

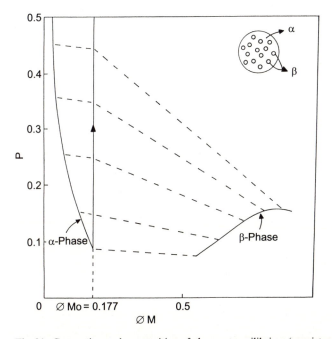

Fig. 21. Conversion and composition of phases at equilibrium (coexistence curves) calculated for an initial volume fraction of modifier, ϕ_{Mo} = 0.177 in a castor oil (CO)-modified DGEBA-EDA system (Reprinted from Polymer, **35**, C.C. Riccardi, J. Borrajo, R.J.J. Williams, Thermodynamic analysis of phase separation in rubber-modified thermosetting polymers: influence of the reactive polymer polydispersity, 5541–5550, Copyright (1994), with kind permission from Butterworth–Heinemann journals, Elsevier Science Ltd, The Boulevard, Langford Lane, Kidlington OX5 1GB, UK)

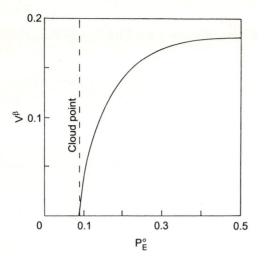

Fig. 22. Volume fraction of dispersed phase (β-phase) as a function of the overall conversion of epoxide groups in the CO-modified DGEBA-EDA system (Reprinted from Polymer, **35**, C.C. Riccardi, J. Borrajo, R.J.J. Williams, Thermodynamic analysis of phase separation in rubber-modified thermosetting polymers: influence of the reactive polymer polydispersity, 5541–5550, Copyright (1994), with kind permission from Butterworth–Heinemann journals, Elsevier Science Ltd, The Boulevard, Langford Lane, Kidlington OX5 1GB, UK)

a short conversion range close to the cloud-point conversion. This is just the consequence of the application of the lever rule in Fig. 21, i.e. it is the result of the shape of coexistence curves. In a practical situation, a sharp decrease in mass transfer rates (i.e. in diffusion coefficients) is expected as conversion increases [24, 69, 102]. Therefore the effect shown in Fig. 22 will be enhanced, i.e. the evolution of the volume fraction of dispersed β phase will be observed in a very short conversion range close to the cloud point.

Figure 23 shows the ratios of amine/epoxy equivalents in both phases. The β-phase is initially enriched in the amine because EDA ($M = 60 \, \mathrm{g\,mol^{-1}}$) has a much lower size than DGEBA ($M = 368 \, \mathrm{g\,mol^{-1}}$). And the only cause of segregation between both phases is the relative size of species because a single interaction parameter was assumed for all the population, including the monomers. However, as conversion increases, the tetra-functional diamine is consumed at a higher rate that the difunctional epoxide. Then, as at high conversions, the free amount of DGEBA gets much higher than the free amount of EDA, and the β-phase shows a higher concentration of epoxy equivalents than amine equivalents. As the volume fraction of β-phase is much lower than that of the α-phase, the stoichiometric imbalance in the latter is practically not observed.

In the next section we will discuss what the situation looks like in a case which is closer to the behaviour of actual systems.

2.2.4 Secondary Phase Separation

Even in the framework of a purely thermodynamic simulation, the effect introduced by mass transfer limitations may be dealt with by considering another

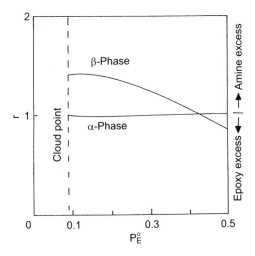

Fig. 23. Ratios (r) of amine epoxy equivalents in both phases as a function of the overall conversion of epoxide groups in the CO-modified DGEBA-EDA system (Reprinted from Polymer, **35**, C.C. Riccardi, J. Borrajo, R.J.J. Williams, Thermodynamic analysis of phase separation in rubber-modified thermosetting polymers: influence of the reactive polymer polydispersity, 5541–5550, Copyright (1994), with kind permission from Butterworth–Heinemann journals, Elsevier Science Ltd, The Boulevard, Langford Lane, Kidlington OX5 1GB, UK)

limiting case. As the β-phase is richer in the (usually) viscous modifier than the α-phase, it will be assumed that epoxy-amine species (including the monomers) remain trapped in the dispersed domains. Therefore the β-phase will receive material from the continuous phase but will not deliver any species back to the α-phase, a situation resembling more the behaviour of actual systems. The β-phase is thus semipermeable to mass transfer. A limit of this situation is when the modifier is a thermoplastic; the β-phase rich in thermoplastic can vitrify if the reaction temperature, T_r is lower than the glass-transition temperature of the β-phase.

In this case, with the β-phase semipermeable to mass transfer, the thermodynamic simulation was carried out as follows. Once α and β-phases were generated, kinetic equations were solved independently for both phases. After a differential time, the α-phase was driven to equilibrium segregating a differential amount of material rich in modifier that was incorporated in the β-phase. At this time, the β-phase, modified both by the material received from the α-phase and the evolution of species through the continuation of the polymerization, was driven to equilibrium. Under these conditions, a secondary phase separation (i.e. a phase separation inside the β-phase) took place, as shown in Fig. 24. The γ-phase (dispersed phase inside particles of the β-phase) is rich in epoxy-amine copolymer whereas the δ-phase (continuous matrix inside particles of the β-phase – also called submatrix), is rich in the modifier. While γ- and δ-phases are always at equilibrium, α- and β-phases are not, due to the semipermeable character of the latter. It is observed that, as most of the phase separation takes place at conversions close to the cloud point, the β-phase keeps a significant proportion of epoxy-amine copolymer even at high overall conversions. This agrees with experimental estimations of the composition of dispersed-phase particles in rubber-modified epoxies [103, 104].

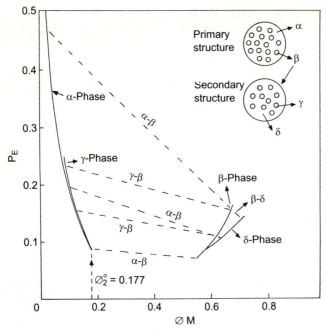

Fig. 24. Coexistence curves when the β-phase is semipermeable to mass transfer. A secondary phase separation is produced inside the β-phase leading to γ and δ phases at equilibrium (Reprinted from Polymer, **35**, C.C. Riccardi, J. Borrajo, R.J.J. Williams, Thermodynamic analysis of phase separation in rubber-modified thermosetting polymers: influence of the reactive polymer polydispersity, 5541–5550, Copyright (1994), with kind permission from Butterworth–Heinemann journals, Elsevier Science Ltd, The Boulevard, Langford Lane, Kidlington OX5 1GB, UK)

Ratios of the amine/epoxy equivalents in different phases are shown in Fig. 25. Now, due to its semipermeable character, the β-phase retains the excess diamine monomer segregated at low conversions, leading to a significant departure from stoichiometry. However, the effect on the α-phase (amine defect) is not so significant because of its high volume fraction in the system.

The occurrence of a secondary phase separation inside dispersed phase particles, associated with the low conversion level of the β-phase when compared to the overall conversion, explains the experimental observation that phase separation is still going on in the system even after gelation or vitrification of the α-phase [26–31]. A similar thermodynamic analysis was performed by Clarke et al. [105], who analyzed the phase behaviour of a linear monodisperse polymer with a branched polydisperse polymer, within the framework of the Flory–Huggins lattice model. The polydispersity of the branched polymer was treated with a power law statistics, cut off at some upper degree of polymerization dependent on conversion and functionality of the starting monomer. Cloud-point and coexistence curves were calculated numerically for various conversions. Spinodal curves were calculated analytically up to the gel point. It was shown that secondary phase separation was not only possible but highly probable, as previously discussed.

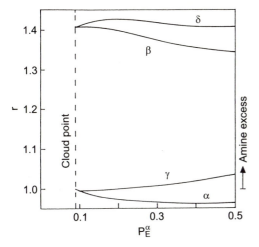

Fig. 25. Ratios (r) of amine epoxy equivalents in every phase as a function of the overall conversion of epoxide groups (Reprinted from Polymer, **35**, C.C. Riccardi, J. Borrajo, R.J.J. Williams, Thermodynamic analysis of phase separation in rubber-modified thermosetting polymers: influence of the reactive polymer polydispersity, 5541–5550, Copyright (1994), with kind permission from Butterworth–Heinemann Journals, Elsevier Science Ltd, The Boulevard, Langford Lane, Kidlington OX5 1GB, UK)

Experimental observations of the phase separation process, in situ by small-angle-X-ray scattering (SAXS) [30, 106], gave clear evidence of the secondary phase separation. The thermosetting polymer was a DGEBA-based epoxy cured with a cycloaliphatic diamine (4,4'-diamino-3,3'-dimethyldicyclohexylmethane, 3DCM), in the presence of 15 wt% of two different modifiers (an epoxy-terminated butadiene-acrylonitrile copolymer, ETBN, and a non-functionalized butadiene-acrylonitrile copolymer, NFBN). The integrated intensity of the scattered radiation [30] was recorded as a function of the polymerization time at 50 °C. Results are shown in Fig. 26.

The particular SAXS technique was sensitive to the presence of a second phase with domains in the range of about 30–500 Å. The peak at short times is related to the development of the primary morphology (segregation of the β-phase from the α-phase); β-phase particles increase their size by a nucleation-growth mechanism and, at a certain time, are too big to be recorded by the SAXS technique. For the ETBN modifier the scattered intensity increases again due to the formation of sub-particles (γ-phase) of about 100 Å diameter by secondary phase separation. The size of γ-phase sub-particles was estimated from SAXS curves using the Guinier method [30] and also directly from transmission electron microscopy (TEM) micrographs. As observed in Fig. 26, the secondary phase separation continues well beyond gelation of the α-phase.

For the NFBN modifier, no increase of the SAXS intensity after the first peak was observed. This was correlated to the fact that the γ-phase, located at the core of subparticles as revealed by TEM, had dimensions greatest than those of the measurement window of the SAXS technique.

Although no generalization can be performed, it is of interest to assess that modifiers with end-groups reacting with the thermosetting polymer [107] lead to a salami structure inside dispersed phase particles, whereas non-functionalized modifiers lead to a core-shell structure, i.e. a rigid core surrounded by an

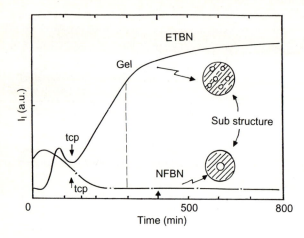

Fig. 26. Integrated SAXS intensity (arbitrary units) vs polymerization time at 50 °C for an epoxy-diamine system (DGEBA-3DCM) modified by 15 wt% of two different rubbers (ETBN and NFBN). Morphology of the particles (β-phase), as observed by transmission electron microscopy (TEM), is plotted qualitatively. Cloud-point times (t_{cp}), determined by light transmission, and gelation times are indicated

elastomeric shell. In this way, the reaction-induced phase separation process offers the unique possibility of producing different sub-structures by an adequate selection of the modifier.

2.2.5 Limitations of the Thermodynamic Approach

The thermodynamic description of the reaction-induced phase separation employed in this section belongs to the second level of approximations in the classification defined by Kleintjens and Koningsveld [55]. Even at this level, several simplifying assumptions have been adopted, for instance a single interaction parameter dependent only on temperature and conversion. Moreover, no considerations were made of the changes introduced in the thermodynamic analysis by the reaction of a functionalized modifier with the thermosetting polymer; a significant driving force to improve the thermodynamic frameworks is thus available.

However, qualitative (and in some cases quantitative) explanations of the most significant experimental observations are possible.

3 Factors that Control the Morphologies Generated

3.1 Nucleation-Growth (NG) and Spinodal Demixing (SD)

When the system evolves in the metastable region (i.e. the region comprised between binodal and spinodal curves in the simple thermodynamic analysis depicted by Fig. 11), the phase separation mechanism is determined by the ratio [107]

$$K = \text{phase separation rate/cure reaction rate.} \qquad (45)$$

Figure 27 shows possible trajectories in the metastable region according to the K value.

If $K \rightarrow \infty$, equilibrium is instantaneously reached and the system evolves along the binodal curve (trajectory a). The segregated phase composition follows the other branch of the binodal curve (trajectory a').

If $K \rightarrow 0$, no phase separation will be observed until the spinodal curve is reached (trajectory c), at which point separation takes place by a continuous and spontaneous process by a diffusional flux against the concentration gradients (uphill diffusion with a negative diffusion coefficient). Morphologies generated display some degree of connectivity (bicontinuous structure). At later stages, coarsening of the structure takes place.

Trajectory b represents the general case in which phase separation takes place by nucleation and growth (followed by coalescence and Ostwald ripening at later stages [108–112], at a rate insufficient to achieve equilibrium conditions. Compositions of the dispersed phase are indicated by trajectory b', which lies outside the metastable region [24, 25, 65]. The location of ϕ_{Mo} with respect to ϕ_{Mcrit} also plays a role in determining the phase separation mechanism. SD is favoured near ϕ_{Mcrit} while NG has a higher probability of occurring at off-critical compositions. A Monte Carlo simulation of the phase separation process in a thermoplastic-modified epoxy [113], showed that SD occurred near the critical point irrespective of the magnitude of the cure reaction rate, but at off-critical compositions the mechanism changed from NG to SD as the cure rate increased.

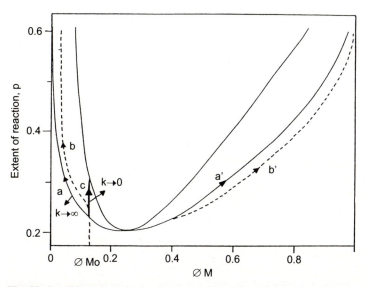

Fig. 27. Possible trajectories in the metastable region of the conversion vs composition phase diagram; a' and b' represent the compositions of the dispersed phase corresponding to trajectories a and b, respectively (K = phase separation rate/cure reaction rate. Trajectories a and b are associated to an NG mechanism while trajectory c leads to SD)

There is significant evidence indicating that phase separation occurs by an NG mechanism when the modifier concentration is lower than and sufficiently apart from the critical concentration (Fig. 19), i.e. in the usual range used in commercial formulations of rubber-modified epoxides.

Direct proofs of this assessment are as follows: a) optical micrographs obtained in situ during the phase separation process of a rubber-modified epoxy showed that, after a certain reaction time, many tiny spots appeared; the average diameter grew gradually with cure time until it was fixed at around 3 μm, well before gelation [114]; b) diffusion coefficients measured in situ during phase separation have been found to be positive [69, 102, 115]; c) in situ scattering tudies using SAXS generated typical spectra expected for a random dispersion of spherical particles (Fig. 28) [30, 106]; d) the use of experimental techniques with different measurement windows, like SAXS, light transmission (LT) and scanning electron microscopy (SEM), shows an increase in the average size of dispersed phase particles during the phase separation of a rubber-modified epoxy (Fig. 29) [30] (the increase in the average size is qualitatively explained by an NG mechanism followed by Ostwald ripening; the primary morphology is settled well before gelation); e) cloud-point curves rather than spinodal curves were obtained by in situ measurements of phase separation by LT (Fig. 19); f) several observed experimental trends could be explained using expressions for the phase separation rate derived from the NG mechanism [24, 25, 65, 107, 115, 116].

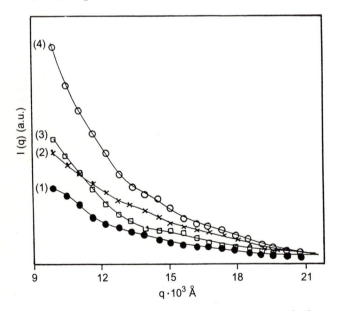

Fig. 28. SAXS intensity as a function of the scattering angle, for a rubber (ETBN) modified epoxy-amine (DGEBA-3DCM) system reacted at 50°C, at different cure times. (1) 80 min, (2) 105 min, (3) 130 min, (4) 180 min (Reprinted from Polymer International, **32**, D. Chen, J.P. Pascault, H. Sautereau, G. Vigier, Rubber-modified epoxies. II. A reaction-induced phase separation observed in-situ and a posteriori with different methods, 369–379, Copyright (1993), with kind permission from the Society of Chemical Industry, London, UK)

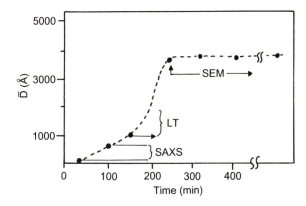

Fig. 29. Several experimental points of the change of particle size during phase separation in a rubber (ETBN) modified epoxy-amine (DGEBA-3DCM) system, determined by different techniques: SAXS (small-angle X-ray scattering), LT (light transmission), SEM (scanning electron microscopy)

Therefore, trajectory b) in Fig. 27 represents the most general situation describing the phase separation process in modified thermosetting polymers at off-critical concentrations.

Spinodal demixing (SD) has been observed when the modifier is a thermoplastic of high molar mass and for modifier concentrations close to or higher than the critical concentration (Fig. 19), and particularly for high cure-rate systems or high interfacial tension between both phases (K → 0) [117–122]. Computer simulations were carried out to evaluate the time-dependent concentration fluctuations during the SD demixing process under non-isoquench depth conditions. (SD driven by reaction proceeds isothermally but the quench depth, ΔT, in terms of the temperature difference between the equilibrium temperature and the reaction temperature, increases with time due to the increase in conversion) [123]. The simulations revealed that the SD under successive increases in quench depth yields a regular two-phase structure as in the case of the SD under isoquench. The structure coarsening is suppressed by the increase in quench depth and the final morphology is highly dependent on the quench rate. The computer simulation adequately describes the characteristic features of structure development in the reaction-induced SD.

It has been argued [118, 122] that the NG mechanism cannot take place at all because homogeneous nucleation is an extremely slow process. However, based on the conclusive experimental evidence now available, the following question is to be asked: how does nucleation take place in spite of being an intrinsically slow process?

Two answers may be provided at this time. The first one deals with the homogeneous nucleation process. The nucleation rate, NR, is proportional to [124]:

$$NR \approx \exp(-\Delta G_c / k_B T) \tag{46}$$

where k_B is the Boltzmann constant and ΔG_c is the free energy barrier for nucleation, which is proportional to the third power of the interfacial tension

(σ_0), i.e.

$$\Delta G_c \approx \sigma_0^3. \tag{47}$$

Therefore the value of σ_0 plays a significant role in starting an homogeneous nucleation process.

Figure 30 shows a simulation of different trajectories in the metastable region, obtained by solving differential equations for polymerization, nucleation and growth rates, using σ_0 as an adjustable parameter [65, 107]. If $\sigma_0 > 0.1$ mN m^{-1}, spinodal decomposition can easily be attained. In modified-thermosetting polymers, phase separation leads to a solution of different composition (i.e. richer in modifier) than the starting solution. Interfacial tensions of demixed polymer solutions derived from polymer-polymer-solvent systems were reported [125, 126] in the range of 10^{-4}–10^{-1} mN m^{-1}. Therefore nucleation of a new phase can easily be attained.

A second answer was provided by Binder [127], who suggested that in polymer systems nucleation is an heterogeneous event, the starting nuclei being the largest species of the polymer molar-mass distribution.

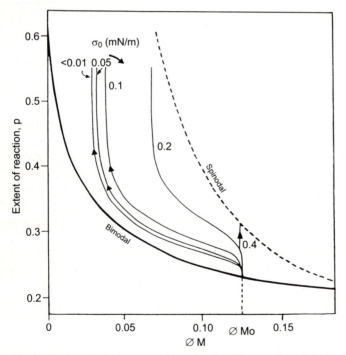

Fig. 30. Trajectories in the metastable region for different values of the interfacial tension (trajectories represent the composition of the continuous phase at different conversion levels). The simulation shows that increasing the interfacial tension leads to a decrease of the phase separation in the metastable region (Reprinted from Journal of Applied Polymer Science, **42**, S.M. Moschiar, C.C. Riccardi, R.J.J. Williams, D. Verchère, H. Sautereau, J.P. Pascault, Rubber-modified epoxies. III. Analysis of experimental trends through a phase separation model, 717–735, Copyright (1991), with kind permission from John Wiley & Sons, Inc., New York, USA)

Regarding the secondary phase separation process, the question whether it occurs by NG or SD may also be asked. As it is unlikely that the composition of the coexisting phases are such that they are close to the critical points of their own secondary phase diagrams, any secondary phase separation will probably be via nucleation and growth [105]. For example, structures of epoxy-polysulfone (70/30) blends obtained by reaction-induced phase separation exhibit a primary spinodal structure which is typical of blends undergoing the primary phase separation in the critical region of their phase diagrams. Structures also exhibit secondary phase-separated nucleated sites within each of the bicontinuous phases [105]. A similar behaviour was observed in a polystyrene/poly(styrene-co-butadiene) (SBR) blend obtained from the chain-polymerization of styrene in presence of SBR as modifier [128].

Returning to the argument that NG is not a significant mechanism in reaction-induced phase separations [118, 122], the direct experimental evidence supporting this statement is the generation of characteristic light-scattering patterns (ring pattern and the presence of a maximum in the intensity of the scattered light as a function of the scattering angle [122]). In this regard, it is interesting to comment on results obtained by Okada et al. [129]. They studied the phase separation induced by chain-polymerization of 2-chlorostyrene in a polystyrene/dibutyl phthalate mixture, followed by a time-resolved light scattering technique (LS) and scanning electron microscopy (SEM). It was found that, at a certain conversion, the scattered intensity started to increase rapidly while the angular dependence exhibited a maximum at a nonzero angle that was shifted to smaller angles with time. This is precisely the behaviour observed in the ordinary spinodal decomposition. However, observations by SEM revealed that the poly(2-chlorostyrene)-rich phase formed droplets, and that droplets of relatively narrow distribution coexisted with much smaller droplets in an early period. Coalescence and coarsening of droplets occurred later leading to different final morphologies depending on cure temperature. Therefore assignation of an SD mechanism on the exclusive basis of the LS behaviour must be made with caution.

In another study, a structure factor was calculated by a Monte Carlo simulation of the phase separation process in a thermoplastic-modified epoxy [113]. The factor could be related to the intensity of the scattered light and plotted as a function of the scattering vector. For conditions favouring an NG mechanism (low reaction rates and off-critical compositions), the appearence of a maximum in the (simulated) scattering light spectra was reported. Increasing the cure time led to an increase in the intensity and a shift of the maximum to lower values of the scattering vector. As time increased, a monotonic decrease of the scattered light as a function of the scattering vector was predicted.

Therefore the most convenient way to proceed in order to establish the phase separation mechanism is to use different experimental techniques giving different size scales of the morphology generated, i.e. SAXS and LS [30], or, when possible, to observe the evolution of morphologies by SEM or TEM at the same cure times as LS observations [129].

In the rest of this section, factors that control the morphologies generated in the phase separation proceeding by an NG mechanism, will be discussed. It is evident that every factor affecting the thermodynamic description (i.e. location of CPC and spinodal curve), the polymerization kinetics and the phase separation rate, should have a bearing on the resulting morphologies. Moreover, varying one parameter of the formulation may vary both thermodynamic and kinetic factors, making it difficult to provide a simple explanation of the observed effect. To make things worse, there is abundant literature in the field where one can find the trend which fits his-her own results (i.e. it has been reported that the volume fraction of dispersed phase remains practically constant [28,103,130,131], goes through a maximum [40] or decreases [40, 41, 76] as temperature increases).

In this context, a qualitative description of the most significant trends will be provided. This will leave aside some particular situations which would also require particular explanations.

3.2 Factors that Control the Location of the Miscibility Gap

The first factor that should be considered when selecting a particular formulation is that the modifier must be expelled as much as possible from the thermosetting matrix. On the one hand, whatever the role of the modifier, it has to be accomplished in the dispersed phase (i.e. toughening by the presence of dispersed-phase particles). On the other hand, leaving modifier dissolved in the matrix may lead to a decrease in thermal properties, i.e. a Tg reduction.

A particular illustration of these concepts is given by the use of three butadiene-acrylonitrile copolymers with number-average molar masses of 2030–3600 and 6050 $g\,mol^{-1}$ to modify an epoxy (DGEBA)-diamine (3DCM) system [72]. Figure 12 shows the location of the miscibility gap for each one of the modifiers, when curing at 50 °C. For an initial volume fraction of modifier, $\phi_{Mo} = 0.177$, the high-molar-mass modifier (H-modifier) exhibits a cloud-pont conversion, $p_{cp} = 0.12$ whereas the medium (M-modifier) and low (L-modifier) molar mass modifiers show p_{cp} values equal to 0.29 and 0.41, respectively. For the H-modifier, the volume fraction of dispersed phase at the end of phase separation was $V_D = 23\%$ and the glass transition temperature of the epoxy matrix (after a postcure to get full conversion), was Tg = 172 °C (compared to 180 °C for the pure matrix devoid of modifier) [72]. Corresponding values for the two other modifiers were: $V_D = 25\%$ and Tg = 154 °C for the M-modifier and $V_D = 15\%$ and Tg = 144 °C for the L-modifier [72]. It is evident that when p_{cp}, at the prescribed cure temperature, approaches p_{gel}, a significant amount of modifier remains dissolved in the matrix, decreasing its Tg value.

Figure 31 shows the qualitative trend between the volume fraction of dispersed phase and the cloud-point conversion. This trend, which has also been predicted by a theoretical simulation of the phase separation process [24], is the

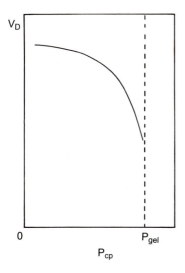

Fig. 31. Qualitative trend between the volume fraction of dispersed phase and the cloud-point conversion

consequence of diffusional limitations to the NG process at high viscosities of the thermosetting matrix. Therefore it is necessary to select the particular modifier so that phase separation takes place in the constant-V_D region of Fig. 31, at the prescribed cure temperature. One can either change the type of modifier (chemical composition, type of end groups, molar mass distribution) or the selected cure cycle (i.e. for a UCST behaviour, decreasing the cure temperature induces a decrease in the cloud-point conversion). In this last option, phase separation and gelation can occur at a convenient low temperature, followed by a postcure at a high temperature to attain full conversion.

The trends discussed in the following sub-sections start from the premise that phase separation proceeds in the constant V_D-region, i.e. there are no diffusional restrictions to complete the phase separation process [28, 103, 130, 131].

3.3 Influence of the Viscosity at the Cloud-Point

Figure 32 [103] shows the viscosity at the cloud point as a function of the average diameter $\bar{D}$ of dispersed phase particles for particular rubber-modified epoxies cured at different temperatures. As the viscosity at the cloud point, η_{cp}, decreases an increase in the average size of dispersed phase particles is observed. Correspondingly, as V_D remains constant, there is a decrease in the concentration of dispersed phase particles, C_{part}, i.e. $\bar{D}$ and C_{part} follow opposite trends as shown in Fig. 33.

Therefore, by changing any variable that affects η_{cp} (i.e. temperature), in the region where V_D remains practically constant, the morphologies generated

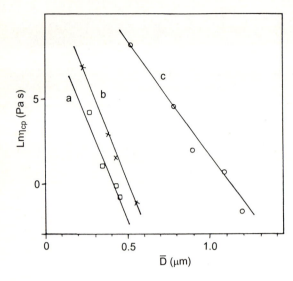

Fig. 32. Viscosity at the cloud point vs average diameter of dispersed phases particules for particular rubber (ETBN) modified epoxies (based on DGEBA) cured at different temperatures a) and b) differ in the nature of the rubber while c) corresponds to a different diamine (Reprinted from Journal of Applied Polymer Science, **42**, D. Verchère, J.P. Pascault, H. Sautereau, S.M. Moschiar, C.C. Riccardi, R.J.J. Williams, Rubber-modified epoxies. II. Influence of the cure schedule and rubber concentration on the generated morphology, 701–716, Copyright (1991), with kind permission from John Wiley & Sons, Inc., New York, USA)

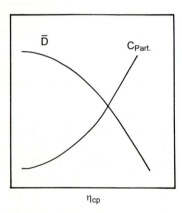

Fig. 33. Average size of dispersed phase particles ($\bar{D}$) and particle concentration (C_{part}) as a function of the viscosity at the cloud point, η_{cp}

change from a low concentration of large particles (at low values of η_{cp}) to a high concentration of small particles (at high values of η_{cp}).

These trends reflect the competition between nucleation, growth, coalescence and Ostwald ripening, particularly through the influence of the diffusion coefficient, which is inversely proportional to the viscosity of the solution [24].

3.4 Influence of Temperature

The influence of reaction temperature on morphologies generated can be inferred from its influence on p_{cp} and η_{cp}. If the temperature range is adequately selected such that V_D is kept constant, i.e. p_{cp} is well below p_{gel}, an increase in temperature will produce a decrease in η_{cp}. In turn, this will provoke an increase in $\bar{D}$ and a decrease in the concentration of dispersed phase particles. Figure 34 illustrates this trend.

3.5 Influence of the Modifier Concentration

The most important factors related to the modifier concentration is the location of ϕ_{Mo} with respect to ϕ_{Mcrit} (composition of the critical point of the CPC). At the left of the critical point, phase separation will produce a dispersed β phase rich in modifier. At the right of the critical point, phase inversion will take place. In a region close to the critical point, it is possible to obtain bicontinuous structures.

Operating with $\phi_{Mo} < \phi_{Mcrit}$, an increase in ϕ_{Mo} will lead to an increase in both the volume fraction of dispersed phase (V_D) and the average size of dispersed phase particles ($\bar{D}$).

3.6 Influence of Reaction Rate

Increasing the polymerization rate by increasing the concentration of a catalyst or initiator will practically affect neither the miscibility (thermodynamics) nor the phase separation rate. Instead, the K factor (Eq. 45) decreases, changing the phase separation mechanism from NG to SD. Thus one should expect a continuous decrease of the volume fraction of dispersed phase and the average size of particles until SD sets in.

In order to eliminate the possibility of setting in SD, a very low modifier concentration should be selected, so that the trajectory could not intercept the spinodal curve. Under these conditions, a continuous decrease of the volume fraction of dispersed phase is expected. This was experimentally verified for a system consisting of diglycidyl ether of resorcinol modified by 3 wt% of

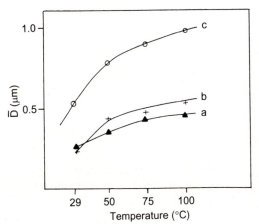

Fig. 34. Average particle size as a function of cure temperature for different rubber (ETBN) modified epoxies (based on DGEBA); a) and b) differ in the nature of the rubber while c) corresponds to a different diamine (Reprinted from Journal of Applied Polymer Science, **42**, D. Verchère, J.P. Pascault, H. Sautereau, S.M. Moschiar, C.C. Riccardi, R.J.J. Williams, Rubber-modified epoxies. II. Influence of the cure schedule and rubber concentration on the generated morophology, 701–716, Copyright (1991), with kind permission from John Wiley & Sons, Inc., New York, USA)

an epoxy-terminated polybutadiene, anionically homopolymerized at 120 °C by the addition of variable amounts of 2,4,6-tri-(dimethylaminomethyl)phenol [115, 132]. Experimental results are shown in Table 1.

Apart from the expected decrease in V_D and $\bar{D}$ when increasing the polymerization rate, at significantly high cure rates bimodal and trimodal particle size distributions are present (Table 1). A quantitative explanation of the appearance of polymodal particle size distributions has been given [116]. When the average distance between growing particles is high and the polymerization rate fast, concentration profiles of modifier may be very sharp, with maxima in points located farthest from any particles. The high supersaturation (concentration excess with respect to equilibrium at the particular conversion level) at these points makes a new nucleation process possible. Different nucleation waves are thus repeated as polymerization proceeds, leading to polymodal particle size distributions.

3.7 Bimodal Particle-Size Distributions

As will be discussed in the following section, there has been an interest in producing bimodal particle-size distributions to improve the toughening effect of the dispersed phase. In Sect. 3.6 it was shown how bimodal (or even polymodal) particle size distributions could be obtained by varying the initiator concentration. However, this procedure is not practical for the different chemistries involved and the range of modifier concentrations used.

Chen and Jan [133] showed that bimodal distributions could be obtained by using two different rubbers as modifiers of a DGEBA-based epoxy resin cured with piperidine. The rubbers were two acrylonitrile-butadiene copolymers (CTBNs), with different AN content, i.e. 18 and 26%. The miscibility with the epoxy resin (and the corresponding cloud-point conversion) increased with the AN content. Therefore, when 10 wt% of CTBN (26% AN) was used as modifier, a high concentration ($C_{part} = 13.4 \ \mu m^{-3}$) of small particles ($\bar{D} = 0.2 \ \mu m$) was obtained. When the same amount of CTBN (18% AN) was used as modifier,

Table 1. Morphology of diglycidyl ether of resorcinol + 3.0 wt% epoxy-terminated polybutadiene + 2,4,6-tri-(dimethylaminomethyl)phenol systems cured at 120 °C

Initial reaction rate, $v_0 \ 10^4$, s^{-1}	Volume fraction of dispersed phase, %	Type of particle size distribution	Average size of particles, $\bar{D}$, μm
0.77	3.0	unimodal	2.5
1.2	2.7	unimodal	1.7
4.1	1.5	unimodal	0.9
6.4	0.6	bimodal	0.55; 0.20
8.3	0.5	trimodal	0.50; 0.17; 0.07
12.1	0.3	bimodal	0.28; 0.07
14.8	0.2	unimodal	0.05
21.0	0.0	–	–

a low concentration ($C_{part.} = 0.032\ \mu m^{-3}$) of large particles ($\bar{D} = 1.9\ \mu m$) resulted. These results are in good agreement with what might be expected, i.e. increasing p_{cp} at constant temperature leads to an increase in and generation of particle size distribution characterized by a high concentration of small particles. The opposite result is observed when decreasing.

When a mixture of both rubbers was used as a modifier, keeping the total amount at 10 wt%, bimodal particle size distributions appeared. When the partial fraction of CTBN (18% AN) increased from 2 to 10 wt% the average diameter of large particles increased from 0.98 to 1.90 μm. The small ones were practically unaffected by variation of the fraction of CTBN (26% AN), and remained at a value close to 0.20 μm. However, their concentration increased significantly from 0.65 to 13.4 μm^{-3} when increasing the CTBN (26% AN) amount from 2 to 10 wt%. Th effect of these bimodal distributions on the fracture toughness of the resulting materials will be discussed in Sect. 4.2.

Another way of producing bimodal particle size distributions is the use of a nucleating agent [134]. A 15 wt% castor oil (CO)-modified epoxy (DGEBA)-amine (EDA) system, was cured at 40 °C for 2 h producing phase separation and gelation. After this time the system was post-cured for another 2 h at 120 °C to get a complete conversion. The resulting morphology was characterized by the following parameters: $V_D = 28.8\%$, $C_{part} = 0.76\ \mu m^{-3}$ and $\bar{D} = 0.75\ \mu m$ (unimodal distribution).

When the 15 wt% modifier was composed of 95 parts CO and 5 parts of a polyester (PE) based on CO and oxalic acid ($\overline{M_n} = 9300\ g\,mol^{-1}$), the high molar mass PE acted as a nucleating agent, and a bimodal particle size distribution was obtained. As the presence of a large molar mass PE shifted the CPC to lower conversions, large particles ($\bar{D} = 3.43\ \mu m$) appear in the population. These particles are present at a very low concentration ($C_{part} = 0.013\ \mu m^{-3}$) but contain most of the volume fraction of dispersed phase ($V_D = 28.0\%$). So PE has been effective as a nucleating agent of the rest of the modifier. However, the low particle concentration sets conditions to increase the supersaturation in regions equidistant from particles [116] making it possible to generate a new wave of nuclei. A peak of small particles ($\bar{D} = 0.16\ \mu m$) appeared in the particle size distribution. These particles exhibit a relatively high concentration ($C_{part} = 0.063\ \mu m^{-3}$) but contain a very small volume fraction of the dispersed phase ($V_D = 0.4\%$).

4 Properties as a Function of Structures Developed

4.1 Influence of Adhesion Between Phases

The interfacial adhesion between dispersed phase particles and the matrix can be improved by functionalizing the modifier with any chemical group that can react with the thermosetting polymer in the selected cure cycle. As has already been discussed, when the end-groups of the modifier are changed a significant shift in its miscibility with the thermoset can be produced. Therefore the particle size and the volume fraction of the rubber-rich phase are modified and it is not possible to vary just the adhesion and keep exactly the same morphologies.

However, using core-shell (CS) particles (rubbery core and rigid shell) dispersed in the initial monomers, the effect of interfacial adhesion can be better analyzed by using comonomers in the shell composition that are capable of reacting with the thermosetting polymer. Sue et al. [49] analyzed this effect by varying the amount of glycidyl methacrylate (GMA) from 0 to 30 wt% shell composition in core-shell particles used to toughen an epoxy-amine matrix. Unfortunately the dispersion of the core-shell particles in the epoxy matrix depends on different parameters (shell composition, shell thickness). In some cases a certain degree of local clustering or agglomeration of CS particles can be observed. In the case of GMA, an appropriate amount of this monomer can help to disperse the CS particles more randomly. Sue et al. concluded that the GMA content played, a minor role, if any, in toughening highly cross-linked epoxy resins. In the absence of GMA, the interfacial strength between the matrix and the core-shell particle is still higher than the cohesive strength of the particle. Therefore cavitation of the particles occurs, followed by shear yielding of the matrix. These are the major toughening mechanisms of these systems.

Huang et al. [135] arrived at a similar conclusion. They stated that once phase separation of the rubber was achieved to give particles of the order of magnitude of micrometers in size, the interfacial bonding between the rubber particles and the matrix had only a small effect on the fracture properties.

For thermoplastic particles dispersed in a thermosetting polymer, the most efficient toughening agents are those that are able to plastically deform and draw by strain hardening (crack bridging mechanism) [136]. Materials which were significantly toughened by thermoplastics possessed good adhesion between the modifier and the matrix material [137]. Without this adhesion, the second phase could prematurely de-bond before deformation could occur. It was hypothesized [136] that if the adhesion were too high, the extent of plastic deformation which the thermoplastic undergoes would be limited due to excessive constraint of the second phase. This constraint would limit the amount of material which could be involved in a bridging and deformation process. Therefore an optimal amount of adhesion must exist.

Experimental studies of the adhesion between thermoset and thermoplastic phases are also of interest. For example, the energy per unit area required to fracture an epoxy-polystyrene interface modified with a carboxylic acid terminated deuterated polystyrene could be increased by as much as 25 times, depending on grafting density and chain length [138].

4.2 Morphologies that Enhance the Fracture Resistance

There is no recipe yet for the desired particle size distribution and composition of dispersed phase particles to give the maximum improvement of the fracture resistance of the thermosetting polymer. The necessity of producing a dispersion of rubber particles with sizes in the range of the micrometers, i.e. from 0.1 to 10 μm, and of producing the maximum volume fraction of dispersed phase without phase inversion are some well-established facts [139]. (It is irrelevant the way in which this distribution of dispersed-phase particles is obtained, i.e. through an NG process or through SD followed by a coarsening of the initial bicontinuous structure.) However the effect of the particle size distribution is still not clear. It has frequently been disappointing to observe that, keeping V_D constant, different particle size distributions gave the same toughening effect [139].

Some indications that bimodal particle size distributions enhance the fracture resistance of toughened epoxy networks have been reported in the literature. Riew et al. [140] showed that the addition of bisphenol A to a CTBN-modified DGEBA cured with piperidine exhibited a bimodal distribution of rubber particles ($\bar{D} = 0.1$ and 1.3 μm) with a higher value of the fracture energy compared to the unmodified formulation. However, as the matrix properties were also changed, it is difficult to generalize this result. Bascom et al. [141] prepared rubber-modified epoxies using two rubbers differing greatly in molar masses. Resulting materials exhibited bimodal particle size distributions with a fracture energy higher than that of networks toughened by either kind of rubber alone. However, the total amount of modifier was not held constant, a fact that masks the net effect of the particles size distribution.

Chen and Jan [133] used two rubbers with different acrylonitrile content to modify an epoxy network cured with piperidine. Bimodal particle size distributions were obtained. By using different proportions of both rubbers and keeping the total amount of modifier in the formulation constant, a synergetic effect on the fracture energy was observed for an intermediate composition (Fig. 35). It has been observed that materials containing one type of rubber exhibit the same fracture energy, even if the particle size distributions are completely different (for 10 wt% CTBN × 9, $\bar{D} = 1.9$ μm, $C_{part} = 0.032$ μm^{-3}; for 10 wt% CTBN × 13, $\bar{D} = 0.2$ μm, $C_{part} = 13.4$ μm^{-3}). All system with bimodal distributions exhibit higher values of fracture energy than those with unimodal ones. At the maximum of fracture energy, the corresponding particle size distribution shows peaks at 0.2 and 1.35 μm. A mechanistic explanation for the synergistic effect of both types of particles was provided [133].

Levita et al. [53] incorporated 10 wt% of a rubber composed of core-shell particles ($\bar{D} = 0.2$ μm) and an amine-terminated butadiene acrylonitrile copolymer (ATBN) in different proportions in a DGEBA-based epoxy resin cured with piperidine at 120 °C for 12 h. The core-shell particles were always immiscible in the epoxy resin while the ATBN phase separated during poly-

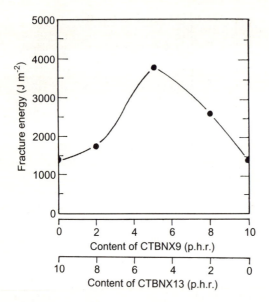

Fig. 35. Dependence of fracture energy on the modifier composition (CTBN 1300×9 = carboxyl-terminated acrylonitrile, acrylic acid and butadiene rubber with 18% acrylonitrile and 2% acrylic acid contents; CTBN 1300×13 = carboxyl-terminated acrylonitrile, butadiene rubber with 26% acrylonitrile content) (Reprinted from Journal of Materials Science, **27**, T.K. Chen, Y.H. Jan, Fracture mechanism of toughened epoxy resin with bimodal rubber-particle size distribution, 111–121, Copyright (1992), with kind permission from Chapman & Hall, London, UK)

merization yielding a second dispersed phase with an average size, $\bar{D} = 5\,\mu m$. Better toughness was obtained for pure core-shell particles and for formulations containing 2/8 parts of ATBN/core-shell particles. The absence of a synergistic effect may be due to the large size of the particles rich in ATBN.

Therefore it seems that the search must continue to find which bimodal distributions are better for toughening purposes.

In the case of thermoplastic-toughened thermosetting polymers, there is ample evidence showing that the fracture energy is significantly increased when bicontinuous or phase-inverted structures are produced [142]. These structures can be obtained either by increasing the thermoplastic content beyond the critical value or by shifting the critical point to lower values as a result of the increase in the thermoplastics molar mass. Some macroscopic studies relating fracture energy to the thermoplastic molar mass do not take into account the possible phase inversion caused by this procedure.

Although it is relatively easy to obtain bicontinuous or phase inverted structures, one should be aware of the fact that other properties, like the solvent resistance, may be deteriorated by the generation of a continuous thermoplastic phase.

4.3 Properties of Thermoset Parts Exhibiting Phase Separation Profiles

One of the advantages of using the reaction-induced phase separation process to generate modified thermosets is that, by an adequate selection of cure cycles and initial formulations, parts exhibiting positionally different morphology can be

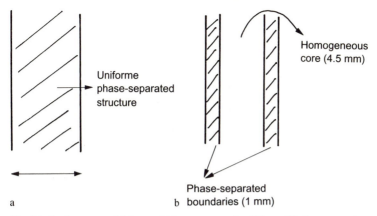

Uniforme phase-separated structure

Homogeneous core (4.5 mm)

Phase-separated
a   b boundaries (1 mm)

Fig. 36a, b. Structure of 6.5 mm-thickness parts of a CO-modified epoxy-amine system, cured in a heated mold: **a** at $0.5\,°C\ min^{-1}$; **b** at $5\,°C\ min^{-1}$ (rates of temperature increase at the boundary)

obtained. Fang et al. [42] showed that a castor oil (CO)-modified epoxy-amine system, cured in a heated mould, could give uniform or non-uniform parts with sandwich structures (Fig. 36) depending on the selected cure cycle. Although both materials exhibited similar values of fracture resistance, the yield stress of the sandwich structure was 86.3 MPa compared to 73.3 MPa for the uniform structure. Due to the fact that the core does not contain a rubber dispersion, the sandwich material also showed a higher elastic modulus than the uniform one.

Therefore it is suggested that modifiers giving special magnetic, electrical or optical properties, differing in solution and dispersion, can be used for an in situ generation of anisotropic materials.

5 Concluding Remarks

In the last two decades, numerous experimental and theoretical studies dealing with reaction-induced phase separation in multiphase polymer systems (mostly porous matrices, toughened plastics, melt processable thermoplastics [143], molecular composites, polymer dispersed liquid crystals, etc.) have been reported. A newcomer in this field should get acquainted with hundreds (possibly thousands) of papers and patents. The intention of this review was to provide a qualitative basis (quantitative occasionally) to rationalize the various factors that must be taken into account to obtain desired morphologies.

In the first section it was shown that: a) the reaction-induced phase separation process is a competitive technique with that using a two-phase dispersion from the beginning of the reaction (there are advantages and disadvantages with both procedures); b) the purpose of modifying a thermosetting polymer may be very different for one or another application (desired morphologies are very different in the toughening of an epoxy network, in the development of magnetic disk coatings, or in preparing a PDLC).

The thermodynamic description of the phase separation process was performed using a Flory–Huggins equation with an interaction parameter depending on temperature and conversion. A simple approach in terms of monodisperse constituents was used to get a qualitative insight into the process. Factors controlling the transition of the system from a homogeneous stable state to a metastable or unstable one as conversion increases, were discussed. The possibility of phase inversion and the fact that the dispersed phase contains a significant fraction of thermosetting polymer were established. A more rigorous approach in terms of polydisperse constituents demonstrated that both the entropic contribution to the free energy of mixing, and a variation of the interaction parameter with conversion, must be taken into account to explain the location of cloud-point curves. The presence of a secondary phase separation taking place in dispersed domains was explained by fractionation of the oligomer species of the thermosetting polymer. As monomers and low-molar mass species exhibit a higher concentration in dispersed domains, polymerization may continue along with a secondary phase separation, even after gelation or vitrification of the matrix.

Factors that control the morphologies generated were discussed on a qualitative basis, focusing on practical applications. Ways to develop bimodal particle-size distributions were considered. These morphologies may enhance the fracture resistance of toughened-materials when compared with unimodal distributions. The possibility of producing phase separation profiles in cured parts was established. Thus reaction-induced phase separation may be used for in situ production of anisotropic materials.

From a general point of view, the challenge in the case of reaction-induced phase separation is to create new morphologies (particle sizes at the nanometer scale, transparent two-phases materials, interconnected phases, etc.) and to control the properties of the interface (adhesion, and also the internal stress concentration). For size adjustment, it is possible to superimpose a thermal quench to the reaction-induced phase separation [144]. In the case of thermoplastic blends the interfaces or interphases can be modified and modelled by the use of block copolymers, especially triblock terpolymers ABC [145]. A similar systematic approach can be developed in thermoset blends and the results are expected to be different to those obtained by the use of functionalized modifiers [71, 138, 146–148].

If some one has chosen to use a two-phase dispersion from the beginning of the reaction, the particle morphology is well-defined and cannot be modified, but swelling of the shell of the dispersed particles by monomers offers the possibility of controlling the properties of the interphase. To prepare dispersed particles with long hairs on their surface to model a wet brush effect dispersed in the reactive system represents another possibility.

The experimental description of the reaction-induced phase separation at different times and scales by the use of different techniques would be useful to understand better the parameters (modifier structure, reaction rate, etc.) which control the phase separation process (NG and/or SD).

At present there is a need to improve our quantitative knowledge of this subject. This would require improvement of the thermodynamic framework (i.e. improvement of the approximation level in the classification suggested by Kleintjens and Koningsveld [55]) and mathematical description of phase separation rates [24], including Ostwald ripening, the presence of concentration gradients [116], possible heterogeneous nucleation [127] or the vitrification of the dispersed phase, etc.). It would be necessary to describe factors affecting final morphologies when phase separation takes place by spinodal demixing [117–123]. Also the sensitivity to shear of the phase-separated morphology was demonstrated in the literature [149]; from the practical (composite processings, moulding compounds, etc.) and scientific points of view, this phenomenon is very important and more experiments are needed in the future.

Within the thermodynamic framework proposed, only one reaction has been taken into account. But, generally, the modifier can also react with the matrix. Besides, the introduction of a second reaction quite different from the first one, for example chain and step reactions, is essential for modelling more complex systems like interpenetrating polymer networks.

Finally, the subject of the present review, the reaction-induced phase separation, is present in several other areas of the polymer field. For example, composition fluctuation effects in chain copolymerization can be sufficient to induce liquid-liquid phase separation or mesophase separation within the reaction medium [150]. This has a bearing in bulk and emulsion polymerizations.

Acknowledgement. Authors wish to thank Prof. K. Dušek (Institute of Macromolecular Chemistry, Prague) for his valuable remarks and discussions as well as for bringing to their attention some of his pioneering publications in the field [34–36, 59].

6 References

1. Huang Y, Hunston DL, Kinloch AJ, Riew CK (1993). In: Riew CK, Kinloch AJ (eds) Toughened plastics 1: science and engineering. Adv Chem Ser 233, Am Chem Soc, Washington DC, p 1
2. Riew CK, Gillham JK (eds) (1984). In: Rubber-modified thermoset resins. Adv Chem Ser 208, Am Chem Soc, Washington DC
3. Riew CK (ed) (1989). In: Rubber toughened plastics. Adv Chem Ser 222, Am Chem Soc, Washington DC
4. Riew CK, Kinloch AJ (1993). In: Riew CK, Kinloch AJ (eds) Toughened plastics 1: science and engineering. Adv Chem Ser 233, Am Chem Soc, Washington DC
5. Sperling LH, Heck CS, An JH (1989). In: Utracki LA, Weiss RA (eds) Multiphase polymers: blends and ionomers. Adv Chem Ser 395, Am Chem Soc, Washington DC
6. Suspène L, Fourquier D, Yang YS (1991). Polymer 32: 1593
7. Hsu CP, Kinkelaar M, Hu P, Lee LJ (1991). Polym Eng Sc 31: 1450
8. Bucknall CB, Davies P Partridge IK (1985). Polymer 26: 109
9. Bucknall CB, Partridge IK, Phillips MJ (1991). Polymer 32: 786
10. Bucknall CB, Partridge IK, Phillips MJ (1991). Polymer 32: 636
11. Lucas JC, Borrajo J, Williams RJJ (1993). Polymer 34: 1886

12. Burns JM, Prime RB, Barrall EM, Oxsen ME, Wright SJ (1989). In: Mittal KL (ed) Polymers in Information Storage Technology. Plenum, New-York, p 237
13. Ho TH, Wang CS (1993). J Appl Polym Sci 50: 477
14. West JL (1988). Mol Cryst Liq Cryst 157: 427
15. Kim JY, Palffy-Muhoray P (1991). Mol Cryst Liq Cryst 203: 93
16. Dušek K (1982). In: Haward RN (ed) Developments in Polymerisation-3. Applied Science, Barking (UK), p 143
17. Mc Garry FJ, Willner AM (1968). Org Coat Plast Chem (ACS) 28: 512
18. Sultan JN, Laible RC, Mc Garry FJ (1971). J Appl Polym Sci 6: 127
19. Sultan JN, McGarry FJ (1973). Polym Eng Sci 13: 29
20. Pearson RA (1993). In: Riew CK, Kinloch AJ (eds) Toughened Plastics 1: Science and Engineering. Adv Chem Ser 233; Washington DC; Am Chem Soc, p 405
21. Kim BS, Chiba T, Inoue T (1993). Polymer 34: 2809
22. Bucknall CB, Gomez CM, Quintard I (1994). Polymer 35: 353
23. Pascault JP, Galy J, Mechin F (1994). In: Hamerton I (ed) Chemistry and Technology of Cyanate Ester Resins. Chapman & Hall, London, p 112
24. Williams RJJ, Borrajo J, Adabbo HE, Rojas AJ (1984). In: Riew CK, Gillham JK (eds) Rubber-Modified Thermoset Resins. Adv Chem Ser 208 Washington DC; Am Chem Soc p 195
25. Vazquez A, Rojas AJ, Adabbo HE, Borrajo J, Williams RJJ (1987). Polymer 28: 1156
26. Visconti S, Marchessault RH (1974). Macromolecules 7: 913
27. Wang TT, Zupko HM (1981). J Appl Polym Sci 26: 2391
28. Chan LC, Gillham JK, Kinloch AJ, Shaw SJ (1984). In: Riew CK, Gillham JK (eds) Rubber-Modified Thermoset Resins. Adv Chem Ser 208 Washington DC; Am Chem Soc p 235
29. Grillet AC, Galy J, Pascault JP (1992). Polymer 33: 34
30. Chen D, Pascault JP, Sautereau H, Vigier G (1993). Polym Int 32: 369
31. Delides CG, Hayward D, Pethrick RA, Vatalis AS (1993). J Appl Polym Sci 47: 2037
32. Seidl J, Malinsky J, Dušek K, Heitz W (1967). Adv Polym Sci 5: 113
33. Kun KA, Kunin R (1968). J Polym Sci Part A-1, 6: 2689
34. Dušek K (1971). In: Chompff A (ed) Polymer networks: Structural and Mechanical Properties. Plenum, New York, p 245
35. Dušek K (1967). J Polym Sci C16: 1289
36. Dušek K (1971). Chem Zvesti 25: 177
37. Moerkerke R, Koningsveld R, Berghmans H, Dušek K, Solč K (1995). Macromolecules 28: 1103
38. Ruseckaite RA, Hu L, Riccardi CC, Williams RJJ (1993). Polym Int 30: 287
39. Enns JB, Gillham JK (1983). J Appl Polym Sci 28: 2567
40. Manzione LT, Gillham JK, Mc Pherson CA (1981). J Appl Polym Sci 26: 889
41. Manzione LT, Gillham JK, Mc Pherson CA (1981). J Appl Polym Sci 26: 907
42. Fang DP, Frontini PM, Riccardi CC, Williams RJJ. Polym Eng Sci 35: 1359
43. Adabbo HE, Williams RJJ (1982). J Appl Polym Sci 27: 1327
44. Williams RJJ (1985). In: Whelan A, Craft J (eds) Developments in Plastics Technology-2. Elsevier, London, p. 339
45. Bucknall CB (1989). In: Allen G and Bevington JC (eds) Comprehensive Polymer Science – The Synthesis, Characterization, Reactions and Applications of Polymers 7, p 27
46. Hoffman DK, Arends CB: US Patent 4,708,996 (The Dow Chemical Co) 1987
47. Hoffman DK, Ortiz C, Hunston DL, Mc Donough W (1994). Polym Mat Sci Eng (ACS) 70: 7
48. Ortiz C, Mc Donough W, Hunston DL, Hoffman DK (1994). Polym Mat Sci Eng (ACS) 70: 9
49. Sue HJ, Garcia-Meitin EI, Pickelman DM, Yang PC (1993). In: Toughened Plastics 1: Science and Engineering. Adv Chem Ser 233; Washington DC; Am Chem Soc, p 259
50. Riew CK, Siebert AR, Smith RW, Fernando M, Kinloch AJ (1994). Polym Mat Sci Eng (ACS) 70: 5
51. Qian JY, Pearson RA, Dimonie VL, El-Aasser MS (1994). Polym Mat Sci Eng (ACS) 70: 17
52. Maazouz A, Sautereau H, Gérard JF (1994). Polym Bull 33: 67
53. Levita G, Marchetti A, Lazzeri A (1991). Makromol Chem, Macromol Symp 41: 179
54. Mülhaupt R, Buchholz U (1994). Polym Mat Sci Eng (ACS) 70: 4
55. Kleintjens LA, Koningsveld R (1988). Makromol Chem, Macromol Symp 20/21: 203
56. Olabisi O, Roberson LM, Shaw MT: Polymer–Polymer Miscibility, New York, Academic Press 1979

57. Koningsveld R, Staverman AJ (1968). J Polym Sci A-2, 6, 305: 325
58. Koningsveld R, Kleintjens LA (1971). Macromolecules 4: 637
59. Dušek K (1969). Coll Czech Chem Comm 34: 3309
60. Qian C, Mumby SJ, Eichinger BE (1991). Macromolecules 24: 1655
61. Mumby SJ, Sher P (1994). Macromolecules 27: 689
62. Tompa H (1956). Polymer solutions, Butterworth, London
63. Dušek K (1972). J Polym Sci, C39: 83
64. Gordon M, Goldsbrough J, Ready B, Derham KW (1973). In: Industrial Polymers, Character-ization by Molecular Weight, Transcripta Books, London, p 45
65. Moschiar SM, Riccardi CC, Williams RJJ, Verchère D, Sautereau H, Pascault JP (1991). J Appl Polym Sci 42: 717
66. Verchère D, Sautereau H, Pascault JP, Moschiar SM, Riccardi CC, Williams RJJ (1989). Polymer 30: 107
67. Roginskaya GF, Volkov VP, Chalykh AE, Avdeev NN, Rozenberg BA (1979). Enikolopyan NS Vysokomol Soedin A21: 2111
68. Chalykh AE, Volkov VP, Roginskaya GF, Avdeev NN, Matveev VV, Rozenberg BA (1981). Plast Massy 4: 25
69. Volkov VP, Roginskaya GF, Chalykh AE, Rozenberg BA (1982). Usp Khim 51: 1733
70. Roginskaya GF, Volkov VP, Kuzaev AI, Chalykh AE, Rozenberg BA (1984). Vysokomol Soedin A26: 1020
71. Chen D, Pascault JP, Bertsch RJ, Drake RS, Siebert AR (1994). J Appl Polym Sci 51: 1959
72. Chen D, Pascault JP, Sautereau H, Ruseckaite RA, Williams RJJ (1994). Polym Int 33: 253
73. Wang HB, Li SJ, Ye JY (1992). Polym Sci 44: 789
74. Iijima T, Tomoi M, Suzuki A, Kakiuchi H (1994). Eur Polym J 27: 851
75. Yorkgitis FM, Tran C, Eiss NS, Hu TY, Yilgor I, Wilkes GL, Mc Grath JE (1984). In: Rubber-Modified Thermoset Resins. Adv Chem Ser 208 Washington DC; Am Chem Soc p 137
76. Seris A, Pascault JP, Camberlin Y (1991). In: Abadie MJM and Sillion B (eds). Polyimides and other High-Temperature Polymers, Elsevier Sci Pub, Amsterdam, p 347
77. Solč K (1974). J Polym Sci Polym Phys Edn 12: 1865
78. Ruseckaite RA, Williams RJJ (1993). Polym Int 30: 11
79. Rehage G, Moller D, Ernst O (1965). Makromol Chem 38: 232
80. Šolc K (1970). Macromolecules 3: 665
81. Kamide K, Miyazaki Y (1981). Polym J 13: 325
82. Kamide K, Abe T, Miyazaki Y (1982). Polym J 14: 355
83. Kamide K, Matsuda S, Dobashi T, Kaneko M (1984). Polym J 16: 839
84. Šolc K, Kleintjens LA, Koningsveld R (1984). Macromolecules 17, 573
85. Matsuda S (1986). Polym J 18: 981
86. Rätzsch MT (1987). Makromol Chem Macromol Symp 12: 101
87. Kamide K, Matsuda S, Shirataki H (1990). Eur Polym J 26: 379
88. Mumby SJ, Sher P, Eichinger BE (1993). Polymer 34: 2540
89. Borrajo J, Riccardi CC, Williams RJJ, Cao ZQ, Pascault JP, Polymer, in press
90. Peebles LH (1971). In: Molecular Weight Distribution in Polymers, New York, Interscience-Wiley
91. Bauer M, Bauer J, Kuhn G (1986). Acta Polym 37: 715
92. Bauer M, Bauer J, Much H (1986). Acta Polym 37: 221
93. Bauer J, Bauer M (1987). Acta Polym 38: 16
94. Stutz H, Simak P (1993). Makromol Chem 194: 3031
95. Fukui K, Yamabe T (1960). J Polym Sci 45: 305
96. Fukui K, Yamabe T (1967). Bull Chem Soc Jpn 40: 2052
97. Shirataki H, Matsuda S, Kamide K (1990). Br Polym J 23: 299
98. Ricardi CC, Adabbo HE, Williams RJJ (1984). J Appl Polym Sci 29: 2481
99. Riccardi CC, Williams RJJ (1987). In: Sedlacek B and Kahovec J (eds) Crosslinked Epoxies, Berlin, de Gruyter, p 291
100. Riccardi CC, Borrajo J (1993). Polym Int 32: 241
101. Riccardi CC, Borrajo J, Williams RJJ (1994). Polymer 35: 5541
102. Roginskaya GF, Volkov VP, Chalykh AE, Matveev VV, Rozenberg BA, Enikolopyan NS (1980). Dokl Akad Nauk USSR 252: 402
103. Verchère D, Pascault JP, Sautereau H, Moschiar SM, Riccardi CC, Williams RJJ (1991), J Appl Polym Sci 42: 701

104. Bussi P, Ishida H (1994). J Polym Sci B: Polym Phys 32: 647
105. Clarke N, Mc Leish TCB, Jenkins SD (1995). Macromolecules 28: 4650
106. Cao ZQ (1994). Doctoral Thesis, INSA de Lyon (France)
107. Verchère D, Sautereau H, Pascault JP, Moschiar SM, Riccardi CC, Williams RJJ (1993). In: Riew CK, Kinloch AJ (eds) Toughened Plastics 1: Science and Engineering. Adv Chem Ser 233; Washington DC; Am Chem Soc, p 335
108. Greenwood GW (1956). Acta Metall 4: 243
109. Lifshitz IM, Slyozov VV (1961). J Phys Chem Solids 19: 35
110. Wagner C (1961). Z Elektrochemie 65: 581
111. Marqusee JA, Ross J (1984). J Chem Phys 80: 536
112. Bartels J, Lembke U, Pascova R, Schmelzer J, Gutzow I (1991). J Non-Cryst Solids 136: 181
113. Jo WH, Ko MB (1994). Macromolecules 27: 7815
114. Chen JP, Lee YD (1995). Polymer 36: 55
115. Rozenberg BA (1991). Makromol Chem, Macromol Symp 41: 165
116. Nikitin OV, Rozenberg BA (1992). Polym Sci USS 34: 365
117. Yamanaka K, Inoue T (1989). Polymer 30: 662
118. Yamanaka K, Takagi V, Inoue T (1989). Polymer 30: 1839
119. Yamanaka K, Inoue T (1990). J Mater Sci 25: 241
120. Hsich HSY (1990). J Mater Sci 25: 1568
121. Hsich HSY (1990). Polym Eng Sci 30: 493
122. Inoue T (1995). Prog Polym Sci 20: 119
123. Ohnaga T, Chen W, Inoue T (1994), Polymer 35: 3774
124. Doremus RH (1985). In: Rates of Phase Transformations. Orlando (FL): Academic
125. Langhammer G, Nester L (1965). Makromol Chem 88: 179
126. Riess G (1986). In Initiation à la Chimie et à la Physicochimie Macromoléculaires. Mélanges des Polymères Vol 6, Strasbourg GFP
127. Binder K (1983). J Chem Phys 79: 6387
128. Donatelli AA, Sperling LH, Thomas DA (1976). Macromolecules 9: 671
129. Okada M, Fujimoto K, Nose T (1995). Macromolecules 28: 1795
130. Butta E, Levita G, Marchetti A, Lazzeri A (1986). Polym Eng Sci 26: 63
131. Montarnal S, Pascault JP, Sauterau H (1989). In: Riew CK (ed) Rubber Toughened Plastics. Adv Chem Ser 222 Washington DC; Am Chem Soc p 193
132. Roginskaya GF, Volkov VP, Dzhavadyan EA, Zaspinok GS, Rozenberg BA, Enikolopyan NS (1986). Dokl Akad Nauk USSR 290: 630
133. Chen TK, Jan YH (1992). J Mater Sci 27: 111
134. Ruseckaite RA, Fasce DP, Williams RJJ (1993). Polym Int 30: 297
135. Huang Y, Kinloch AJ, Bertsch RJ, Siebert AR (1993). In: Riew CK and Kinloch AJ (eds) Rubber Toughened Plastics. Adv Chem Ser 233 Washington DC; Am Chem Soc p 189
136. Cardwell BJ, Yee AF (1994). Polym Mat Sci Eng (ACS) 70: 254
137. Wilkinson SP, Ward TC, Mc Grath JE (1993). Polymer 34: 870
138. Norton LJ, Smigolova V, Pralle MU, Hubenko A, Dai KH, Kramer EJ, Hahn S, Berglund C, Dekoven B (1995). Macromolecules 28: 1999
139. Verchére D, Pascault JP, Sautereau H, Moschiar SM, Riccardi CC, Williams RJJ (1991). J Appl Polym Sci 43: 293
140. Riew CK, Rowe EH, Siebert AR (1976). In: Deanin RD and Crugnola AM (eds) Tougheness and Brittleness of Plastics. Adv Chem Ser 154 Washington DC Am Chem Soc p 326
141. Bascom WD, Ting RY, Moutton RJ, Riew CK, Siebert AR (1981). J Mater Sci 16: 2657
142. Kinloch AJ, Yuen ML, Jenkins SD (1994). J Mater Sci 29: 3781
143. Venderbosch RW, Nelissen JGL, Meijer HEM, Lemstra PJ (1993). Makromol Chem Macromol Symp 75: 73
144. de Graaf LA, Möller M (1994). Makromol Chem Macromol Symp 77: 149
145. Auschra C, Stadler R (1993). Macromolecules 26: 6364:
146. Hendrick JL, Yilgor I, Wilkes GL, Mv Grath JE (1985). Polym Bull 13: 201
147. Cecere JA, Mc Grath JE (1986). Polym Prepr (Am Chem Soc Div Polym Chem) 27(1): 299
148. Hendrick JC, Patel NM, Mc Grath JE, in Ref 4 (1993). chap 11, 293
149. Van Dijk A, Eleveld MB, Van Veelen A (1992). Macromolecules 25: 2274
150. Fredrickson GH, Leibler L (1995). Macromolecules 28: 5198

Editor: K. Dusek
Received: October 1995

The Physics of Polymer Dissolution: Modeling Approaches and Experimental Behavior

Balaji Narasimhan and Nikolaos A. Peppas
School of Chemical Engineering, Purdue University, West Lafayette,
IN 47907-1283, USA

Polymer dissolution is an important phenomenon in polymer science and engineering that has found applications in areas like microlithography, controlled drug delivery, and plastics recycling. This review focuses on the modeling efforts to understand the physics of the dissolution mechanism of glassy polymers. A brief review of the experimentally observed dissolution behavior is presented, thus motivating the modeling of the mechanism of dissolution. The main modeling contributions have been classified into four broad approaches – phenomenological models and Fickian equations, external mass transfer-control based models, stress relaxation models, and anomalous transport models and scaling law-based approaches. Another approach discussed is the appropriate accommodation of molecular theories in a continuum framework. The underlying principles and the important features of each approach are discussed in depth. Details of the important models and their corresponding predictions are provided. Experimental results seem to be qualitatively consistent with the present picture.

Advances in Polymer Science, Vol. 128
© Springer-Verlag Berlin Heidelberg 1997

List of Symbols and Abbreviations

Roman Symbols

A	Constant of Eq. (71) dependent on polymer molecular weight, solvent viscosity and temperature
a	Primitive path step length
a_d	Exponent in the free volume diffusivity model, Eq. (69)
B	Constant of Eq. (77) dependent on polymer molecular weight, solvent viscosity and temperature
b	Bond length
b_i	Body force of component i
c	Polymer mass concentration
D	Macroscopic diffusion coefficient
D_{coop}	Cooperative diffusion coefficient
D_p	Stokes-Einstein diffusion coefficient
D_R	Rouse diffusion coefficient
D_{self}	Self diffusion coefficient
D_0	Preexponential factor in free volume diffusivity model, Eq. (69)
D_1	Free volume diffusion coefficient
D_{12}	Mutual diffusion coefficient
D_2	Reptation diffusion coefficient
E	Modulus in Maxwell element
F	Deformation gradient tensor
F(t)	Fraction of polymer present in original tube
f_g	Free volume fraction of polymer
f_1	Numerical factor for the diffusion coefficient of solvent in polymer, Eq. (1)
f_2	Numerical factor for diffusion coefficient of dissolved polymer in liquid solution, Eq. (2)
ΔG	Total change in Gibbs free energy
ΔG_E	Change in Gibbs free energy due to elastic expansion
ΔG_M	Change in Gibbs free energy due to mixing
ΔG_{seg}^{OR}	Orientational contribution to the Gibbs free energy of a segment
g	Volume fraction of polymer in an entanglement subunit
ΔH	Enthalpy change during elastic expansion
j_i	Diffusional flux of component i
K	Parameter of kinetic model for glass transition, Eq. (21)
k	Boltzmann constant
k_d	Disentanglement rate
k_1	Mass transfer coefficient
L	Parameter of Eq. (32) dependent on c (or π)
L(t)	Average primitive path length
l	Half thickness of thin polymer slab
M_c	Critical molecular weight of polymer

M_e	Molecular weight between entanglements
$\bar{M}_n$	Number averaged molecular weight of the polymer
m_p	Mobility of disengaging polymer chain
$m_{p,\infty}$	Maximum mobility of disengaging polymer chain
N	Number of repeating units
N_e	Number of moles of entanglements
n	Parameter of kinetic model for glass transition, Eq. (21)
n_i	Number of moles of component i
Pe	Peclet number
p_i^+	Exchange of linear momentum between components
R	Position of glassy-rubbery interface
R_d	Dissolution rate
R_{eff}	Effective disengagement rate
R_0	Radius of the polymer particle
r	Radial position
r_g	Radius of gyration
S	Position of rubbery-solvent interface
ΔS	Entropy change
Sh	Sherwood number
T	Temperature
T_g	Glass transition temperature
t	Time
t_d	Disentanglement time
t_{rep}	Reptation time
U_R	Reference velocity scale, Eq. (16)
U_∞	Velocity of solvent stream
V_m	Monomer volume
V_1	Average volume of solvent molecule
$\bar{V}$	Molar volume of the solvent
V_2	Average volume of a polymer chain
$\bar{V}_2$	Molar volume of the polymer
v	Local swelling rate of the polymer
v_1	Convective velocity of the solvent in the x-direction
x	Position
$\bar{x}_n$	Ratio of polymer molar volume to solvent molar volume
Z	Number of segments in the primitive path

Greek Symbols

α	Isotropic expansion factor
β	Scaling factor in expression for disentanglement time, Eq. (59)
γ	Constant for critical stress level, Eq. (48)
δ	Thickness of diffusion boundary layer
ε	One-dimensional deformation

η	Viscosity
θ	Overall dissolution time
κ	Constant appearing in Eq. (24)
μ_i	Chemical potential of component i
μ_i^0	Chemical potential of pure component i
μ_1^{OP}	Chemical potential of solvent due to osmotic pressure
μ_1^{OR}	Chemical potential of solvent due to orientational contribution
V_{eff}	Number of entanglements per polymer chain
ξ	Distance between entanglements
π	Osmotic pressure
ρ_i	Density of component i
σ	Stress in the rubbery polymer
σ_c	Critical stress for crazing
σ_{xx}	Normal stress component of the stress tensor
τ_{dif}	Characteristic diffusion time
υ_d	Equilibrium solubility of polymer in solvent
υ_i	Volume fraction of component i
υ_1^*	Critical solvent volume fraction, at which mode of mobility changes
$\upsilon_{1,eq}$	Equilibrium solvent volume fraction
$\upsilon_{1,t}$	Threshold solvent volume fraction for swelling
$\upsilon_{2,b}$	Polymer volume fraction in the bulk liquid
Φ	Factor that determines the extent of local swelling, Eq. (37)
χ	Polymer-solvent interaction parameter

1 Introduction

1.1 The Polymer Dissolution Problem

The dissolution of a polymer in a penetrant involves two transport processes, namely penetration of the solvent into the polymer, followed by disentanglement of the macromolecular chains. When an uncrosslinked, amorphous, glassy polymer is in contact with a thermodynamically compatible liquid (solvent), the latter diffuses into the polymer. A gel-like layer is formed adjacent to the solvent-polymer interface due to plasticization of the polymer by the solvent. After an induction time, the polymer is dissolved. A schematic diagram of solvent diffusion and polymer dissolution is shown in Fig. 1. However, there also exist cases where a polymer cracks when placed in a solvent.

The main features of the dissolution mechanism of a polymer are depicted as shown in Fig. 2. During the initial stage of the dissolution process, a glassy polymer of thickness 2L starts swelling due to the penetration of the solvent into it and the simultaneous transition from the glassy to the rubbery state. Thus, two distinct fronts are observed – a swelling interface at position R and a polymer/solvent interface at position S. Front R moves inwards while front S moves outwards. When the concentration of the penetrant in the polymer exceeds a critical value, macromolecular disentanglement begins and true dissolution commences. After an induction time, the polymer is dissolved. During this time, front R continues to move towards the center of the slab, while front S moves inwards as well. After the disappearance of the glassy core, only front S exists and continues to move inwards towards the center of the slab till all of the polymer is dissolved.

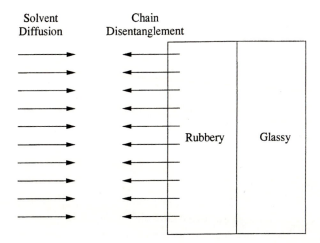

Fig. 1. A schematic diagram of one-dimensional solvent diffusion and polymer dissolution

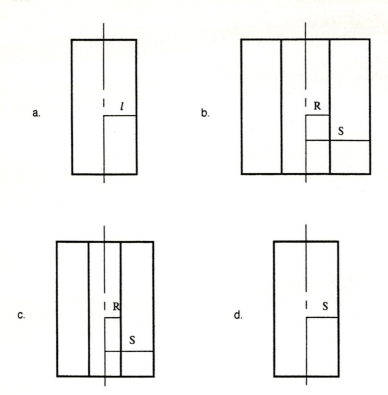

Fig. 2a–d. Schematic representation of a one-dimensional solvent diffusion and polymer dissolution process: **a** initial slab of thickness 2L: **b** initial swelling step showing the increasing position of the rubbery-solvent interface (S) and the decreasing position of the glassy-rubbery interface (R): **c** onset of the dissolution step showing the decreasing position of the interface S along with the decreasing position of the interface R: **d** final dissolution step where the slab has been transformed into a rubbery material (disappearance of interface R) and the position of interface S still decreases

In general, polymer dissolution differs from dissolution of a non-polymeric material in two aspects. Polymers require an induction time before starting to dissolve, while non-polymeric materials dissolve instantaneously. Also, polymer dissolution can be controlled either by the disentanglement of the polymer chains or by the diffusion of the chains through a boundary layer adjacent to the solvent-polymer interface. However, the dissolution of non-polymeric materials is generally controlled by the external mass transfer resistance through a liquid layer adjacent to the solid-liquid interface.

1.2 Applications of Polymer Dissolution

Dissolution of a polymer in solvents is an important phenomenon in polymer science and engineering. It has found significant applications in a variety of areas. For example, in microlithography, a process used in fabricating integrated circuits, a photosensitive polymer or photoresist, is coated onto a substrate to form a thin film [1–5]. Irradiation through a glass plate or "mask" bearing an array of circuit patterns allows selected areas of the photoresist to be exposed (Fig. 3). The modified or exposed regions of the polymer exhibit an altered rate of removal in certain solvents, resulting in the formation of a polymeric image of the mask pattern.

On the basis of the chemical nature of the photoresist, either a positive or a negative image of the original mask is formed. Resists that produce a negative-tone image undergo crosslinking upon irradiation. Crosslinking renders these resists less soluble in the developer solvent. Conversely, positive resists undergo molecular changes that enhance their solubility in the developer solvent so that exposed regions are preferentially removed. The resist image thus obtained delineates the areas in which sequential modification or removal of the underlying substrate will take place. For small devices, as in large scale integrated circuits, resolution of the pattern image becomes important. Thus, the dissolution process has to be controlled carefully so as to minimize solvent swelling of the undissolved polymer and to prevent polymer films from cracking.

In controlled release applications of polymers, a solute is molecularly dispersed in a polymer phase, which is usually in the glassy state. In the presence of a thermodynamically compatible solvent, swelling occurs and the polymer begins to release its contents to the surrounding fluid. This release process can

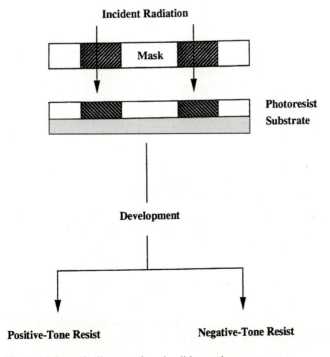

Fig. 3. A schematic diagram of a microlithography process

be controlled either by solvent diffusion or by polymer dissolution. In this case, the presence of permanent entanglements in the polymer becomes significant. These systems are often called erosion-controlled release systems and can give constant release rates under special conditions [6–11].

Polymer dissolution has found applications in membrane science. In phase inversion, a technique to form asymmetric membranes, a thin film of polymer solution is cast onto a suitable substrate followed by immersion in a coagulation bath (quench step) [12–14]. During the quench period, solvent/non-solvent exchange and eventual polymer precipitation occur. The extent of dissolution of the polymer is instrumental in determining the ultimate structure of the membrane.

Polymer dissolution has also been applied in the treatment of unsorted plastics for recycling. For example, in one such technique recently reported by Nauman and Lynch [15, 16], a single solvent such as xylene is used to dissolve five polymers – poly (vinyl chloride), polystyrene, low density polyethylene, high density polyethylene and polypropylene – at five different temperatures, ranging from room temperature to 138 °C. The remaining poly(ethylene terephthalate) can be dissolved using a separate solvent. In this process, xylene is placed in a chip-filled vat first at room temperature. The polystyrene is dissolved while the other five are not. The xylene solution is then drained to a separate part of the system where it is heated under pressure to about 250 °C. The solution is then sent to a vacuum chamber to undergo flash devolatilization, causing xylene to vaporize instantaneously, and leaving behind pure polystyrene. The same xylene is then sent back to dissolve another polymer at a different temperature and the process continues with the other polymers.

The dissolution of Novolak resins in solvents is an important process in many semiconductor applications [17, 18]. Hence, it becomes necessary to understand the dissolution characteristics of Novolaks. Novolak resins, as the world's oldest synthetic materials, play a remarkable role in today's semiconductor industry. Their advantages are in their non-swelling nature, aqueous-base developability and etching resistance [19, 20].

In addition to the above, polymer dissolution rate data have been used to determine glass transition temperature and other thermodynamic parameters associated with polymorphic changes [21]. Dissolution has also found a variety of uses in the pharmaceutical sciences. In the development of microcapsules for sustained release dosage forms [22], the mechanism of drug transport is governed by the dissolution of the polymer. Cooney [23] studied the dissolution of pharmaceutical tablets in the design of sustained release forms. Ozturk et al. [24, 25] showed that the dissolution of the polyacid, which is used in enteric-coated tablets, was the controlling step in the release kinetics mechanism.

1.3 Polymer Dissolution Behavior

We will now summarize the important results of various experimental studies that have contributed to the knowledge of the dissolution behavior of polymers.

Ueberreiter [26] was one of the earliest researchers to study polymer dissolution behavior. He summarized the structure of the surface layer of glassy polymers during dissolution, recognizing

 (i) a liquid boundary layer, surrounding the solid polymer;

 (ii) a gel layer;

(iii) a solid swollen layer; and

(iv) an infiltration layer.

He also proposed two types of dissolution. If the polymer permitted the existence of all four layers described above, the dissolution was termed "normal". The main feature of this type of dissolution was the presence of a gel layer. The other type of dissolution was one without a gel layer. In this case, cracks were observed running into the polymer matrix and these caused small blocks of the polymer to erode and leave the surface.

By lowering the experimental temperature, transition from normal dissolution to the type of dissolution without a gel layer was observed in the dissolution of poly(methyl methacrylate) (PMMA) [27]. The temperature at which the gel layer disappeared was termed as the gel temperature. Ueberreiter and Asmussen [28] observed that polystyrene (PS) undergoes normal dissolution in most solvents owing to its low gel temperature.

Ueberreiter [26] also studied the effect of polymer molecular weight on the dissolution rate and on the gel layer thickness. For PS dissolution in toluene, the dissolution rate varied inversely with the polymer molecular weight over a molecular weight range between 1000 and 200 000. Also, when the ratio of the gel layer thickness to the polymer radius of gyration was plotted against the logarithm of the polymer molecular weight, a horizontal line was obtained over a molecular weight range between 1000 and 100 000.

Ouano and Carothers [29] used a critical angle illumination microscopy technique to study in situ dissolution of PS and PMMA. They observed that in methyl ethyl ketone (MEK), polystyrene developed a thick swollen layer while PMMA cracked. They suggested that the difference in free volume and the segmental stiffness might be responsible for the difference in behavior. They also studied the dissolution of PMMA in tetrahydrofuran, methyl acetate and methyl isobutyl ketone (MIBK). On the basis of their experimental observations, they suggested that the cracks appear sooner in the case of smaller molecular weight solvents than in the bulkier ones due to the higher diffusion rates. Also, the cracks were observed to appear sooner in thermodynamically compatible solvents than in poor solvents. This was attributed to the swelling capacity of the thermodynamically compatible solvent molecules. They pointed out that polymer morphology significantly affects dissolution.

Polymers have been used as positive photoresists in microlithographic applications. Rodriguez and his coworkers [1, 30–33] measured the dissolution rate of PMMA in MEK using laser interferometry. When the logarithm of the dissolution rate was plotted against the logarithm of the polymer molecular weight, over a molecular weight range from 6000 to 320 000, nonlinear behavior was observed. Plasticization of PMMA by poly(ethylene oxide) (PEO) of molecular weight 4000 changed the dissolution rate in direct proportion to the amount of PEO added. Cooper et al. [31] found that the addition of small amounts of low molecular weight non-solvents such as water, methanol, ethanol, 1-propanol, 2-propanol and ethylene glycol increased the rate obtained with a high molecular weight solvent. Groele and Rodriguez [34] investigated the effect of polymer composition on the dissolution rate. They studied homopolymers of methyl methacrylate (MMA), ethyl methacrylate (EMA), n-butyl methacrylate (BMA) as well as copolymers of MMA with EMA and BMA. The polymer dissolution rate in MIBK at 30 °C varied from 0.042 µm/min (PMMA) to more than 150 µm/min (PBMA), showing that copolymers of MMA with EMA and BMA dissolved more rapidly than PMMA. They attempted to explain these observations due to the thermodynamic compatibility of the copolymers with MIBK and the T_g of the copolymers. Recently, Rao et al. [35, 36] measured the dissolution rate of diazonaphthaquinone resists using laser interferometry and showed that the spatial distribution of the sensitizers affects the dissolution mechanism.

Dissolution of PMMA in different solvents was also investigated by Bell and coworkers [37–41]. The effect of polymer molecular weight on the dissolution rate and the gel layer thickness was studied by ellipsometry. It was observed [38] that the dissolution rate in MIBK was inversely proportional to the polymer molecular weight up to a molecular weight of 100 000. The rate leveled off at higher molecular weights. The explanation provided by the authors was that the dissolution rate was controlled by solvent penetration. The penetration was in turn controlled by either polymer relaxation or crazing, depending upon polymer molecular weight. At low molecular weights, the critical stress for crazing was a function of polymer molecular weight. Hence, both solvent diffusion and dissolution mechanisms depended upon the polymer molecular weight. Specifically, the critical stress for crazing increased with increasing polymer molecular weight upto a molecular weight of about 100 000 and then it leveled off. This implied that higher molecular weight polymers could support higher stress levels before the onset of crazing.

The effect of polymer molecular weight on the thickness of the swollen gel layer was also studied [39]. For all ketones, the degree of swelling was minimal. When MIBK was used, no swollen layer was observed for monodisperse polymer with molecular weight of 72 000. The absence of the gel layer indicated that the dissolution occurred by stress cracking, which was consistent with the dependence of the dissolution rate on polymer molecular weight. For molecular weights greater than 72 000, the thickness of the swollen gel layer increased with increasing polymer molecular weight. This was explained by chain disentanglement. In general, the penetration layer of the solvent, which relates directly to the thickness of the swollen surface layer, must be sufficient to free chains that are anchored in the glassy region. A chain cannot disentangle if a part of it is anchored in the unpenetrated, glassy polymer. However, a completely solvated chain readily disentangles. Higher molecular weight chains are anchored deeper into the polymer, and consequently, the thickness of the swollen layer must be greater to release them.

Peppas and coworkers [4, 5, 42] studied the dissolution characteristics of PMMA and its copolymers. The effects of copolymer composition, nature of solvent and the dose of irradiation on the dissolution rate as well as on the thickness of the gel layer were examined. The results showed that, in most cases, the gel layer thickness increased at the beginning of the dissolution process and then decreased after reaching a maximum.

In addition, important knowledge about the dissolution of polymers can be obtained from pharmaceutical and controlled release literature where polymer swelling and dissolution are studied in conjunction with drug release and dissolution. For example, Harland et al. [9] investigated the swelling and dissolution behavior of a system containing a drug and a polymer. It was shown that the dissolution was characterized by two distinct fronts, one separating the solvent from the rubbery polymer and the other separating the rubbery from the glassy polymer. The drug release was related to a diffusional term (with a $t^{0.5}$ dependence) and a dissolution term (with a t dependence). When the two front movements were synchronized, the drug release rate was independent of time.

Various mathematical models have been proposed to model the dissolution mechanism and to explain the experimentally observed dissolution behavior. We shall now describe in detail the various modeling approaches to understand polymer dissolution.

The approaches to model polymer dissolution can be broadly classified (Table 1) as

(i) use of phenomenology and models with Fickian equations;

Table 1. Publications modeling polymer dissolution

Approach	Description	Reference
Phenomenological	Fickian equations, disassociation rate, moving boundary problem, experiments	[43]
Phenomenological	Fickian equations, disengagement dynamics, external velocity field, experiments	[46–48]
External Mass Transfer	Fickian equations, external mass transfer controlled dissolution, experiments	[51]
Stress Relaxation	Swelling of polymer, viscous yield of network, reptation	[53]
Stress Relaxation	Free energy and chemical potential considerations, reptation	[55]
Anomalous Transport and Reptation	Anomalous transport, disentanglement rate, scaling laws, experiments	[57]
Anomalous Transport and Reptation	Anomalous transport, dissolution clock, scaling laws, experiments	[59]
Molecular Theories in a Continuum Framework	Anomalous transport, concentration field, mobility changes, viscoelastic properties, molecular theories	[61, 65]

(ii) models with external mass transfer as the controlling resistance to dissolution;
(iii) models based on stress relaxation; and
(iv) analysis using anomalous transport models for solvent transport and scaling laws for actual polymer dissolution.

2 Phenomenological Models

2.1 The Multi-Phase Stefan Problem

The model proposed by Tu and Ouano [43] for polymer dissolution assumes Fickian solvent penetration into the polymer. The polymer dissolution problem was modeled as a multi-phase Stefan problem [44]. The key parameter in this model was the disassociation rate, R_d, which was defined as the rate at which the polymer transformed from a gel-like phase to a solution. It was proposed that the dissolution process was "disassociation"-controlled if the polymer diffusion rate in a liquid layer adjacent to the solvent-polymer interface was faster than the disassociation rate, or "diffusion"-controlled if the diffusion rate was slower than the disassociation rate.

The solvent diffusional flux was expressed as

$$j_1 = - D_1 f_1 (\upsilon_2) \frac{\partial \upsilon_1}{\partial x} \quad x < S(t) \tag{1}$$

where D_1 is the diffusion coefficient of the solvent, υ_1 and υ_2 are the volume fractions of solvent and polymer, respectively, $f_1(\upsilon_2)$ is a function of υ_2 and $S(t)$ is the position of the liquid-gel interface.

It was also derived that the polymer diffusional flux across a surface liquid layer ($x > S(t)$) was

$$j_2 = - D_2 f_2 (\upsilon_2) \frac{\partial \upsilon_2}{\partial x} \quad \text{for } x > S(t) \tag{2}$$

The swelling velocity was expressed as

$$v_1 = - j_1 (x, t) \tag{3}$$

The interface position was derived as

$$\frac{dS}{dt} = v_1 - R_d \tag{4}$$

The disassociation rate, R_d, was assumed to be equal to the diffusion of the dissolved polymer into the bulk liquid, i.e.,

$$R_d = j_2 (x, t) \tag{5}$$

Consequently, the governing solvent diffusion equation in the polymer in the region between the liquid-gel interface, S and the external polymer thickness, L, is

$$\frac{\partial \upsilon_2}{\partial t} = D_1 \frac{\partial}{\partial x}\left[f_1(\upsilon_2)\, \upsilon_2\, \frac{\partial \upsilon_2}{\partial x} \right] \quad \text{for } S < x < L \tag{6}$$

The governing polymer diffusion equation in the liquid boundary layer was derived as

$$\frac{\partial \upsilon_2}{\partial t} + \frac{dS}{dt}\frac{\partial \upsilon_2}{\partial x} = D_2 \frac{\partial}{\partial x}\left[f_2(\upsilon_2)\, \frac{\partial \upsilon_2}{\partial x} \right] \quad \text{for } x < S(t) \tag{7}$$

The equation for the interface position, S, was

$$\frac{dS}{dt} = D_2\, f_2(\upsilon_2)\, \frac{\partial \upsilon_2}{\partial x} - D_1\, f_1(\upsilon_2)\, \frac{\partial \upsilon_1}{\partial x} \tag{8}$$

Equations (6) and (7) were solved with two sets of boundary conditions. The first set was "source limited", i.e., disassociation rate-controlled and the second was "flux limited", i.e., the concentration at the interface S was equal to an equilibrium value. The functions f_1 and f_2 were assumed to be unity, i.e., concentration-independent diffusion coefficients were used. The multi-phase Stefan problem was solved numerically [44] using a Crank-Nicholson scheme and the predictions were compared to experimental data for PS dissolution in MEK [45]. Critical angle illumination microscopy was used to measure the positions of the moving boundaries as a function of time and reasonably good agreement was obtained between the data and the model predictions (Fig. 4).

The simulations predict both diffusion-controlled dissolution (Fig. 5) and disassociation-controlled dissolution (Fig. 6). Different forms of the functions f_1 and f_2 were used in the simulations and it was shown that the concentration dependence of the diffusion coefficient is very crucial. The disassociation rate, R_d, was treated as a model parameter and the simulations failed to yield more insight into the actual nature of that rate.

2.2 The Disengagement Dynamics Approach

Devotta et al. [46] considered the dissolution of a polymeric particle with spherical geometry. The glassy-rubbery kinetics was assumed to be rapid. The solvent transport was described through a Fickian equation as

$$\frac{\partial \upsilon_1}{\partial t} = \frac{1}{r^2}\frac{\partial}{\partial r}\left(r^2 D_1 \frac{\partial \upsilon_1}{\partial r} \right) - \frac{1}{r^2}\frac{\partial}{\partial r}\left(r^2 v \upsilon_1 \right) \quad t > 0,\, 0 < r < S(t) \tag{9}$$

where D_1 is the solvent diffusion coefficient, υ_1 is the volume fraction of the solvent in the polymer, $S(t)$ is the moving gel-liquid interface, r the radial

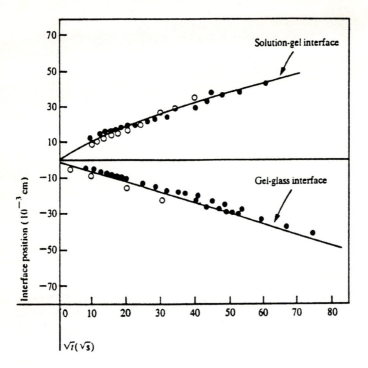

Fig. 4. Positions of the rubbery-solvent and the glassy-rubbery interfaces plotted as a function of the square root of time. The *open circles* are calculated from the model; the *solid circles* are experimental values. Reprinted with permission

position and v(r, t) the local velocity with which the coordinates move. The initial and boundary conditions for the above equations are

$$t = 0, \; 0 < r < R_0 \quad \upsilon_1 = 0 \tag{10}$$

$$r = S(t), \; t > 0 \quad \upsilon_1 = \upsilon_{1,eq} \tag{11}$$

$$r = 0, \; t > 0 \quad \frac{\partial \upsilon_1}{\partial r} = 0 \tag{12}$$

where R_0 is the initial radius of the polymeric particle and $\upsilon_{1,eq}$ the equilibrium volume fraction of the solvent in the polymer.

The position of the moving boundary is given by

$$\frac{dS}{dt} = \left(D_1 \frac{\partial \upsilon_1}{\partial r} \right)_{r=S^-(t)} - \left(\frac{D_p}{\upsilon_{2,eq}} \frac{\partial \upsilon_2}{\partial r} \right)_{r=S^+(t)} \tag{13}$$

where D_p is the diffusivity of the polymer in the solvent and $\upsilon_{2,eq}$ is the equilibrium volume fraction of polymer in the solvent.

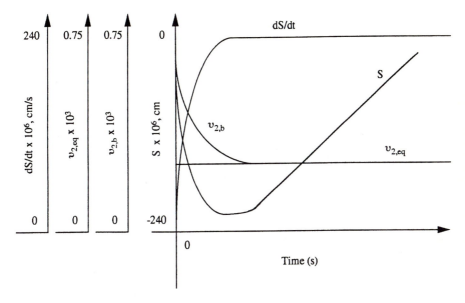

Fig. 5. The rate of removal of dissolved polymer at the boundary layer (dS/dt in cm/s); the liquid phase polymer concentration $v_{2,eq}$, at the rubbery-solvent interface; the polymer concentration at the impermeable surface $v_{2,b}$; and the location of the rubbery-solvent interface, (S), plotted as a function of time during the dissolution of a rubbery polymer. Theoretical predicitions have been adapted from the work of Tu and Ouano [43], using the following parameters: $L_0 = 10^{-4}$ cm; $D_1 = 3 \times 10^{-6}$ cm²/s; $R_d = 10^{-3}$ cm/s; and $v_{2,eq} = 0.25$. This is a diffusion-limited case

To model the chain disengagement, the following boundary condition was written on the liquid side of the liquid-gel interface:

$$r = S^+(t), \quad 0 < t < t_{rep} \quad -D_p \frac{\partial v_2}{\partial r} = 0$$

$$r = S^+(t), \quad t > t_{rep} \quad -D_p \frac{\partial v_2}{\partial r} = k_d \tag{14}$$

where k_d was defined as the disengagement rate. For transport in the liquid, mass transfer was assumed to be the dominating resistance and hence at the liquid side of the liquid-gel interface, as long as $v_2 = v_{2,eq}$,

$$-D_p \frac{\partial v_2}{\partial r} = k_1 (v_{2,eq} - v_{2,b}) \tag{15}$$

where k_1 is the liquid side mass transfer coefficient and $v_{2,b}$ is the polymer volume fraction in the bulk. Correlations were used to obtain the mass transfer coefficient. The model equations were made dimensionless by suitable transforms and solved using a Crank-Nicholson scheme.

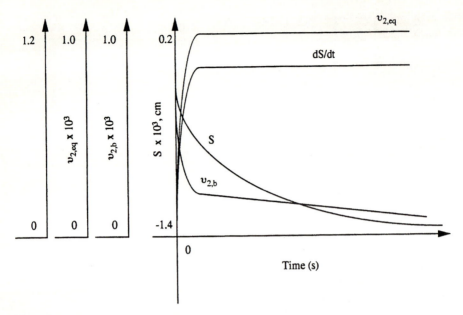

Fig. 6. An example of diassociation-limited dissolution. The flux at the boundary layer; the liquid phase polymer concentration, $v_{2,eq}$; the polymer concentration at the impermeable surface $v_{2,b}$; and the location of the rubbery-solvent interface, S are plotted as a function of dissolution time for the dissolution of a rubbery polymer. Theoretical predictions have been adapted from the work of Tu and Ouano [43], using the following parameters: $L_0 = 10^{-4}$ cm; $D_1 = 3 \times 10^{-6}$ cm^2/s; $D_2 = 10^{-7}$ cm^2/s; $R_d = 10^{-6}$ cm/s; and $v_{2,eq} = 0.05$

The effect of the polymer diffusivity, D_p, on the dissolution time was investigated. It was shown (Fig. 7) that as D_p increases, the dissolution time decreases due to an enhanced mass transfer rate. Also, as the disengagement rate increases, the dissolution time decreases as shown in Fig. 8.

Gravimetric studies were performed with a polystyrene-cyclohexane system. It was observed that there exists a critical particle size of the polymer, below which the dissolution time does not change. This critical size increased with an increase in the rate of stirring as shown in Fig. 9. The predicted dissolution times were compared to the actual dissolution times for PS dissolution in cyclohexane and reasonably good agreement was obtained as presented in Fig. 10.

The above work was extended by Ranade and Mashelkar [47] by considering the dissolution of a spherical polymeric particle in a convective field. The transport equations written were very similar to the ones used by Devotta et al. [46]. In addition, the solvent velocity, U_1, was given as

$$U_1 = \frac{1}{Pe_R}\left(D_p \frac{\partial v_2}{\partial r} + D_1 \frac{\partial v_1}{\partial r}\right) \qquad (16)$$

where Pe_R is a reference Peclet number and D_p and D_1 are the diffusion coefficients of the polymer and the solvent in the gel phase, respectively. Once

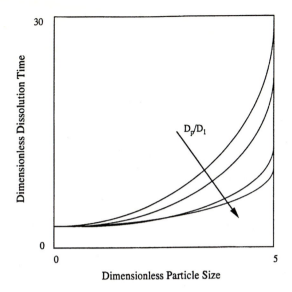

Fig. 7. Effect of the parameter D_p/D_1 on the dissolution time for different particle dimensions. Theoretical predictions have been obtained from the work of Devotta et al. [46]

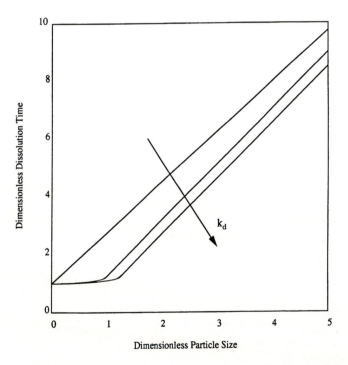

Fig. 8. Effect of the disentanglement rate on the dissolution time for different particle dimensions. Theoretical predictions have been obtained from the work of Devotta et al. [46]

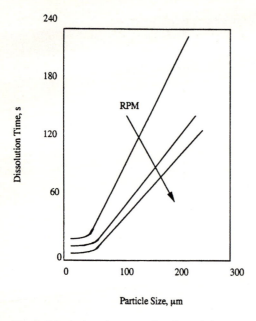

Fig. 9. Dissolution time as a function of the average particle size of polystyrene particles in cyclohexane at different stirring speeds. Theoretical predictions have been abtained from the work of Devotta et al. [46]

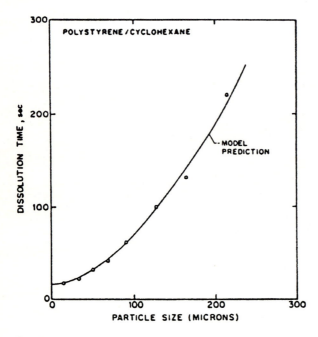

Fig. 10. Comparison of the model predictions with the experimental data for polystyrene-cyclohexane system. Reprinted with permission

again, the boundary conditions written were similar to the ones used by Devotta et al. [46] and the effective disengagement rate, R_{eff}, was treated as a model parameter, similar to the approach used by Tu and Ouano [43].

The dissolving polymer was assumed to be in a uniform solvent stream, moving with a velocity, U_{∞}. Thus, the r component of the mass balance (Eq. (9)) now becomes

$$\alpha \frac{\partial \upsilon_2}{\partial t} + \frac{1}{r^2} \frac{\partial}{\partial r} (r^2 U \upsilon_2) + \frac{1}{r \sin \theta} \frac{\partial}{\partial \theta} (V \sin \theta \, \upsilon_2) =$$

$$\frac{1}{Pe_R} \left\{ \frac{1}{r^2} \frac{\partial}{\partial r} \left(r^2 D_p \frac{\partial \upsilon_2}{\partial r} \right) + \frac{1}{r^2 \sin \theta} \frac{\partial}{\partial \theta} \left(D_p \sin \theta \frac{\partial \upsilon_2}{\partial \theta} \right) \right\} \qquad (17)$$

where α is the ratio of the reference length scale to the product of the reference time and the reference velocity scales, U is the r-component of the velocity and V is the θ-component of the velocity. A creeping flow assumption is made and the external velocity field is determined as

$$U = \left[1 - 1.5 \left(\frac{r}{S} \right) + 0.5 \left(\frac{r}{S} \right)^3 \right] \cos \theta + \left(\frac{S}{r} \right)^2 \frac{dS}{dt} \qquad (18)$$

$$V = - \left[1 - 0.75 \left(\frac{r}{S} \right) - 0.25 \left(\frac{r}{S} \right) \right] \sin \theta \qquad (19)$$

where dS/dt is calculated from Eq. (13).

The simulations yielded the positions of the fronts in the problem as a function of time (Fig. 11) as well as the polymer volume fraction profiles as a function of both time and position (Fig. 12). An increase in the external

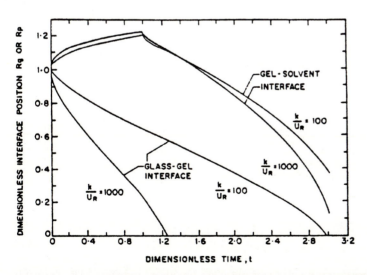

Fig. 11. Positions of the glassy-rubbery and the rubbery-solvent interfaces as a function of time. The parameters are $D_p/D_1 = 0.2$; $R_0 = 2$; $k_d = 5$; Pe = 5000; and $\upsilon_{2,b} = 0.5$. Reprinted with permission

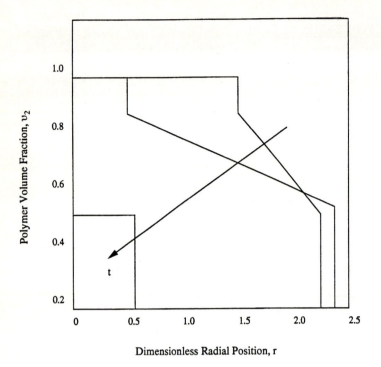

Dimensionless Radial Position, r

Fig. 12. Radial profiles of the polymer volume fraction during dissolution. Theoretical predictions have been adapted from the work of Ranade and Mashelkar [47], using the following parameters: $k_d/U_R = 1000$; $D_p/D_1 = 0.2$; $R_0 = 2$; $k_d = 5$; $Pe = 5000$; and $v_{2,b} = 0.5$

velocity of the solvent, U_∞, caused a decrease in dissolution time. For increased external velocities, it was observed that the dissolution time varied linearly with particle size as shown in Fig. 13. Several limiting cases of the problem were studied by Ranade and Mashelkar [47]. Analytical solutions were presented for cases when the glass transition controlled the dissolution mechanism, the disengagement controlled the process, the external mass transfer controlled the dissolution, and when reptation dominated the mechanism of dissolution. These are presented in Table 2. Some comparisons were made with experiments performed with the PMMA-benzene system and good agreements were obtained as shown in Fig. 14.

This approach was further modified by Mashelkar and his co-workers [48] by considering time dependent "mobility". The relevant transport equation written was

$$\frac{\partial v_2}{\partial t} = \frac{\partial}{\partial x}\left(D_1\, v_2\, \frac{\partial v_2}{\partial x}\right) \quad \text{for } R < x < S \tag{20}$$

where R represents the position of the glassy-rubbery interface. The free volume model of Zielinski and Duda [49] was used to model the concentration

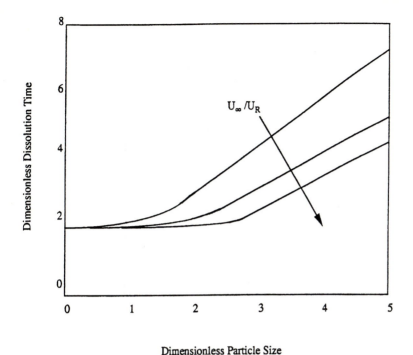

Fig. 13. Effect of external velocity on the dissolution time for different particle sizes. Theoretical predictions have been adapted from the work of Ranade and Mashelkar [47]

dependence of the diffusion coefficient. The kinetics of glass transition was described by a concentration-specified gradient [50] as

$$\frac{dR}{dt} = K(\upsilon_1|_{x=R} - \upsilon_{1,t})^n \tag{21}$$

where $\upsilon_{1,t}$ is a threshold solvent volume fraction for swelling. The terms K and n are experimentally determined. Appropriate initial and boundary conditions were written at $x = R$ and $x = S$. As the polymer chains disengaged, they diffused through a diffusion boundary layer and the transport was modeled as

$$\frac{\partial \upsilon_2}{\partial t} = \frac{\partial}{\partial x}\left[D_p \frac{\partial \upsilon_2}{\partial x}\right] - \frac{dS}{dt}\frac{\partial \upsilon_2}{\partial x} \tag{22}$$

where D_p is the diffusion coefficient of the polymer in the solvent. Once again, appropriate initial and boundary conditions were written by assuming that no polymer disengaged for a time equal to the reptation time of the chain and after a time equal to the reptation time elapses, the chains disengage at a finite rate,

Table 2. Various controlling regimes in polymer dissolution

Controlling Phenomenon	Dissolution Time
Glass Transition	$\dfrac{\alpha R_0}{K \ (\upsilon_{2,t} - \upsilon_{2,b})^n}$
Chain Disengagement	$t_{rep} + \dfrac{\alpha \ (R_d)_{t=t_{rep}}}{k_d}$
External Mass Transfer	$t_{rep} + \dfrac{R_0}{b \ \upsilon_{2,b}^{1/3}} + \dfrac{a}{b} \ln \dfrac{a}{a + \dfrac{R_0}{\upsilon_{2,b}^{1/3}}}$
Chain Reptation	t_{rep}

where $a = \dfrac{0.5 \ D_p^{1/3}}{(R^3 \ g \ \Delta\rho / \mu \ D_R^{1/3})}$ and $b = \upsilon_{2,b} \ D_p^{2/3} \left(\dfrac{D_R^2 \ g \ \Delta\rho}{\mu \ U_R^3}\right)^{1/3}$

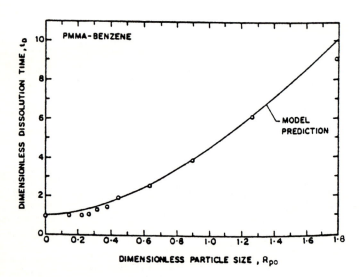

Fig. 14. Experimental data for poly (methyl methacrylate) $(\bar{M}_w = 82\,000)$ – benzene system vs model predictions. Reprinted with permission

k_d. The rubbery-solvent interface moves as a result of solvent transport and polymer disengagement, and is described by

$$\frac{dS}{dt} = \left(D_1 \ \frac{\partial\upsilon_2}{\partial x}\right)_{x=S-} - \left(\frac{D_p}{\upsilon_{2,\,eq}} \ \frac{\partial\upsilon_2}{\partial x}\right)_{x=S+} \tag{23}$$

where $\upsilon_{2,eq}$ is the equilibrium polymer fraction at the rubbery-solvent interface. It was proposed that the disengagement rate was proportional to the mobility of the disengaging polymer chain, m_p. The rate of change of mobility was assumed to be given by a product of a kinetic constant and the extent of departure from the maximum mobility. A free volume dependence of the mobility on the polymer volume fraction was also proposed, given by

$$m_{p,\infty} = \kappa A_d \exp\left(-\frac{B_d}{f_g}\right) \tag{24}$$

where $m_{p,\infty}$ is the maximum mobility, κ is a constant, B_d is a parameter that depends on the polymer molecular weight, and f_g is the "free volume" fraction of the rubber. The concentration at the rubbery-polymer was calculated by using thermodynamics of swollen networks.

The simulations yielded the polymer volume fraction profile as a function of time, both in the rubbery and in the diffusion boundary layer, as shown in Figs. 15 and 16. The variation of the positions of the various interfaces in the problem as a function of time is shown in Fig. 17. The effect of molecular weight

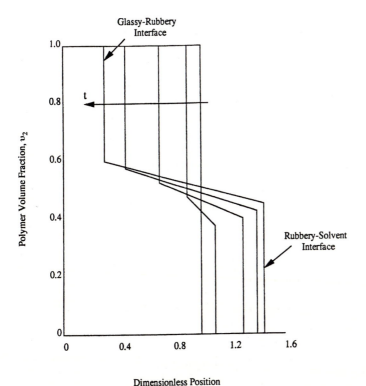

Fig. 15. Polymer concentration profile in the slab, expressed as the polymer volume fraction, v_2, as a function of dimensionless position with 0 indicating the center of the slab. Theoretical predictions have been adapted from the work of Devotta et al. [48], using the following parameters: $v_{1,t} = 0.15$; $B_d = 1.5$; $v_2^* = 0.3$; $\chi = 0.35$; and $KL_0^2/D_0 = 10^{-7}$

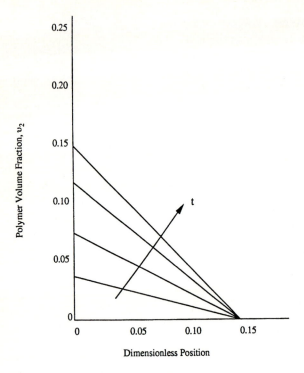

Fig. 16. Polymer concentration profile in the boundary layer, expressed as the polymer volume fraction, υ_2, as a function of dimensionless position with 0 indicating the rubbery-solvent interface. Theoretical predictions have been adapted from the work of Devotta et al. [48], using the following parameters: $B_d = 1.8$; $D_p/D_0 = 5 \times 10^{-4}$; $\delta/L_0 = 0.15$; $KL_0^2/D_0 = 10^{-7}$; $\chi = 0.35$; and $\upsilon_2^* = 0.25$

on the dissolution mechanism was studied by varying the parameter B_d. On decreasing B_d (decreasing molecular weight), a plateau was observed in the variation of the fraction dissolved with time (see Fig. 18). There was no plateau for high enough B_d, which was indicative of disentanglement-controlled dissolution. The effect of the diffusion boundary layer thickness on the dissolution mechanism was also studied. It was observed that, for increased thicknesses, the mechanism shifted from a disentanglement-controlled one to a diffusion-controlled one as shown in Fig. 19.

3 The External Mass Transfer Argument

In dissolution of non-polymeric materials in solvents, the resistance offered by the film adjacent to the material assumes significance. It is hence intuitive to examine the role of such an external mass transfer resistance in polymer dissolution.

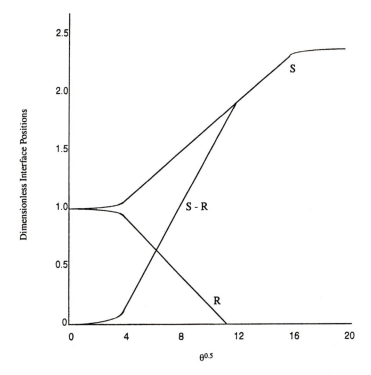

Fig. 17. Variation of the position of the glassy-rubbery interface, the rubbery-solvent interface, and the gel layer thickness as a function of the square root of time. Theoretical predictions have been adapted from the work of Devotta et al. [48], using the following parameters: $\upsilon_{1,t} = 0.15$; $KL_0^2/D_0 = 10^{-7}$; and $\chi = 0.35$

The role of the resistance presented by the liquid film adjacent to a dissolving polymer sample was first investigated by Lee and Peppas [51]. The solvent transport was expressed by a Fickian equation.

$$\frac{\partial \upsilon_1}{\partial t} = \frac{\partial}{\partial x}\left[D_1^v \frac{\partial \upsilon_1}{\partial x}\right] \tag{25}$$

where D_1^v is the volume-based diffusion coefficient. At the glassy-rubbery boundary, R, the interface position is determined by

$$-D_1^v \frac{\partial \upsilon_1}{\partial x} = \upsilon_1 \frac{dR}{dt} \tag{26}$$

For the gel-liquid boundary, S, the interface position is determined by

$$D_1^v \frac{\partial \upsilon_1}{\partial x} - k\upsilon_d = (\upsilon_1 + \upsilon_2) \frac{dS}{dt} \tag{27}$$

where υ_d is the equilibrium polymer volume fraction at the front S, and k is a mass transfer coefficient for the polymer.

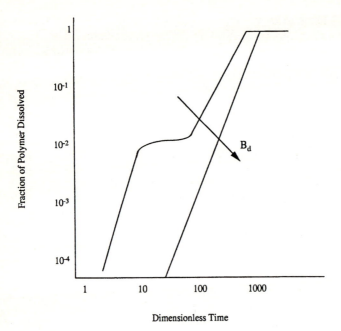

Fig. 18. Effect of the parameter B_d on the dissolution kinetics. Theoretical predictions have been adapted from the work of Devotta et al. [48], using the following parameters: $\delta/L_0 = 0.15$; $D_p/D_0 = 10^{-4}$; $\upsilon_2^* = 0.4$; $KL_0^2/D_0 = 10^{-7}$; and $\upsilon_{1,t} = 0.2$

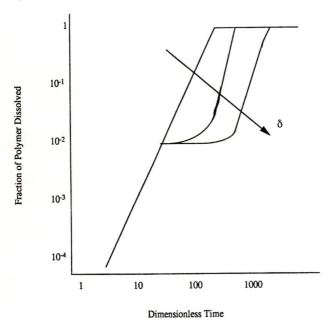

Fig. 19. Effect of thickness of the boundary layer on the dissolution kinetics. Theoretical predictions have been adapted from the work of Devotta et al. [48]. $B_d = 1.5$ and all the other parameters are as in Fig. 18

A quasi-steady state solution of Eqs. (25)–(27) yielded the following square root of time dependence of the normalized gel layer thickness, δ:

$$\delta = \sqrt{\frac{2\ D_1^y\ (v_1^* - v_d)\ (2 - v_1^*)}{L^2(1 - v_1^*)}\,t} \tag{28}$$

Here, v_1^* is the solvent volume fraction at which the glassy-rubbery transition occurred and L is the half-thickness of the polymer slab.

The model predictions were compared to experimental data reported in the literature [5, 43]. The predicted normalized gel layer thickness as a function of the square root of time was compared to experimental values (Figs. 20 and 21) and good agreement was obtained. Comparisons were also made with experimental data for the release of drugs from the polymeric matrix [51].

Lee and Lee [52] also proposed a model for polymer dissolution based on external mass transfer control and obtained a square root of time dependence of the gel layer thickness. This was exactly the same result obtained earlier by Lee and Peppas [51]. Lee and Lee [52] showed that dissolution became important only at large experimental times. This is consistent with the idea that macromolecular chains need a finite time to disentangle.

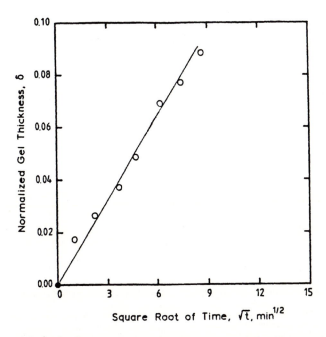

Fig. 20. Normalized gel layer thickness as a function of the square root of methyl ethyl ketone diffusion time for dissolution of polystyrene; data of Tu and Ouano [43]. Reprinted with permission

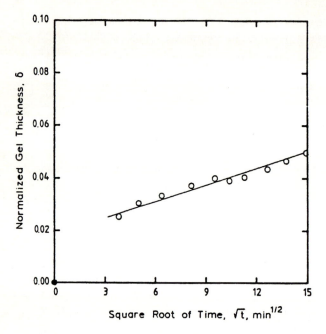

Fig. 21. Normalized gel layer thickness as a function of the square root of water diffusion time for dissolution of drug loaded poly(vinyl alcohol) – poly(n-vinyl 2-pyrrolidone). Reprinted with permission

4 Stress Relaxation and Molecular Theories

The swelling of a polymer owing to solvent transport results in a stress opposing the polymer swelling. Based on this concept, models have been proposed for polymer dissolution.

4.1 The Kinetics of Dissolution

Brochard and de Gennes [53] proposed a relaxation-controlled model for the dissolution of polymer droplets. When a droplet of polymer solution, of concentration v_0, is immersed in a solvent, two processes control the dissolution. The first step relates to the swelling of the polymer network by the solvent. This was assumed to be controlled by the cooperative diffusion coefficient, D_{coop}. The second step corresponds to the viscous yield of the network and is controlled by the reptation time of the polymer, t_{rep}. The expression for the net solvent flux in such a system is given by

$$j_2 = K \frac{\partial}{\partial x} (\sigma - \pi) \tag{29}$$

where K is a proportionality constant dependent on the polymer concentration, c, σ is the stress of the network, and π is the osmotic pressure.

A conservation law was written as

$$\frac{\partial \upsilon_2}{\partial t} + \frac{\partial j_2}{\partial x} = 0 \tag{30}$$

where j_2 is the solvent flux. The osmotic pressure, π, was expressed as a scaling law with respect to the polymer concentration c:

$$\pi \propto c^{2.25} \tag{31}$$

Thus, Eq. (29) is transformed into

$$\frac{\partial \pi}{\partial t} = \frac{9\pi}{4c} \frac{\partial c}{\partial t} = L \frac{\partial c}{\partial t} \tag{32}$$

where L is a function of c (or π). For times much less than the reptation time, the solution of Eq. (31) coupled with an appropriate constitutive equation yields

$$\Delta x = \text{constant} \ (D_{coop} \ t)^{0.5} \tag{33}$$

In other words, the cooperative diffusion coefficient controls the swelling of the transient network, leading to a broadening of the droplet-solvent interface as $(D_{coop} \ t)^{0.5}$.

At finite times, the network yields under the presence of a stress, σ. Using a constitutive function with an appropriate memory function [54], Brochard and de Gennes showed that the viscous yield occurred in a finite time of the order of the reptation time of the polymer. The important parameter arising out of this analysis is l, defined as

$$1 = (D_{coop} \ t_{rep})^{0.5} \tag{34}$$

They further theorized that for large droplets (r(0) > l), the dissolution was limited by the first step, while for small droplets (r(0) < l), the dissolution is limited by the viscous yield. They showed that the optimal size of droplets for rapid dissolution is r(0) $\sim$ l.

4.2 The Reptation Model

Herman and Edwards [55] extended the Brochard and de Gennes approach [53] by considering in detail the stress accompanying the swelling of the polymer within the reptation model. They evaluated the contributions to the free energy and chemical potentials due to the deformation of the polymer due to swelling. The chemical potential of the solvent, μ_1, was obtained by taking two contributions into account. The first was the classical osmotic pressure term, leading to

$$\mu_1^{OP} = -V_1 \pi \tag{35}$$

where V_1 is the average volume of a solvent molecule and π is the osmotic pressure. The second contribution arose from the deformation of the polymer due to solvent swelling. This orientational contribution to the chemical potential was evaluated using reptation theory and is given by

$$\mu_1^{OR} = -\frac{V_1}{V_2} \upsilon_2^2 \left[\frac{dZ}{d\upsilon_2} \Delta G_{seg}^{OR} + Z \frac{\partial}{\partial \upsilon_2} \Delta G_{seg}^{OR} \right] \tag{36}$$

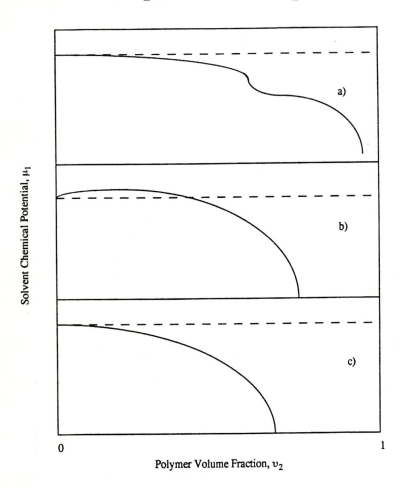

Fig. 22a–c. Solvent chemical potential, μ_1, as a function of polymer volume fraction, υ_2. Theoretical predictions have been adapted from the work of Herman and Edwards [55], using the following parameters: **a** m = 8.0; Z_1 = 0.1; and B = 0.064; **b** m = 1.0; Z_1 = 0.05; and B = 0.26; **c** m = 1.0; Z_1 = 0.01; and B = 2.6. The degree of polymerization is 10 000 in each case. The *dotted line* specifies the zero of the chemical potential. The scale of the chemical potential axis is determined by the choice of the constant kTV_1/V_2, which is left unspecified

where Z is the number of segments in the primitive path and ΔG_{seg}^{OR} is the orientational contribution to the free energy, given by

$$\Delta G_{seg}^{OR} = kT \left[-2 \ln \Phi - 3 + \frac{3 \, \Phi \cos^{-1} \Phi}{\sqrt{1 - \Phi^2}} \right] \tag{37}$$

where Φ is a factor that determines the extent of the local swelling. The net chemical potential is plotted as a function of polymer volume fraction for different cases as shown in Fig. 22. The parameter B appearing in the figure

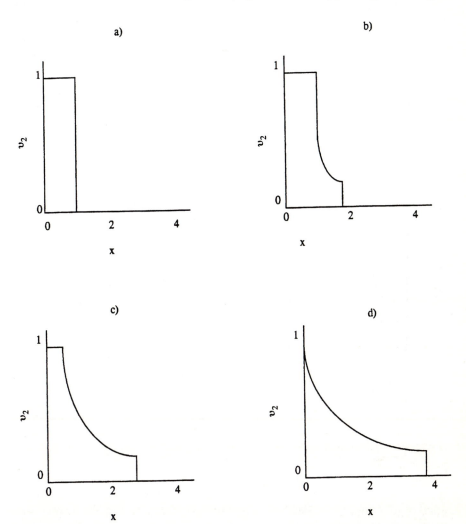

Fig. 23a–d. Polymer volume fraction as a function of dimensionless position for various times. Theoretical predictions have been adapted from the work of Herman and Edwards [55], using the following parameters: $a_d = 10$; and $v_{2,b} = 0.2$. The dimensionless times are: **a** $t = 9.08 \times 10^{-5}$; **b** $t = 4.54 \times 10^{-4}$; **c** $t = 1.36 \times 10^{-3}$; **d** $t = 3.63 \times 10^{-3}$

is given by

$$B = \frac{b^3}{V_m} \tag{38}$$

where V_m is the monomer volume and b is the length of the monomer. At sufficiently long times, the orientational contribution to μ_1 is reduced due to relaxation of the polymer chains, and any phase boundaries, if present, disappear. These ideas were incorporated into simple Fickian transport equations with a free volume diffusion coefficient. The flux boundary condition at $x = S$ was written as

$$\left[-D \frac{\partial \upsilon_2}{\partial x} \right]_{x=S} = \upsilon_{2, b} \frac{dS}{dt} \tag{39}$$

where $\upsilon_{2,b}$ is the polymer volume fraction in the bulk. The transport equations were made dimensionless and numerically solved. A typical polymer volume fraction v/s position profile is shown in Fig. 23. It is seen that υ_2 gradually approaches the bulk value for long times. The phase boundaries disappear when this happens.

5 Anomalous Transport Models and Scaling Laws

Diffusion anomalies are intimately associated with the polymer's glassy/rubbery state transition. Hence, anomalous transport models have been proposed to understand penetration into polymers.

5.1 The Scaling Approach

Recognizing the presence of entanglements in polymers, dissolution has been understood as the transformation undergone by the polymer chains from an entangled gel-like phase to a disentangled liquid solution [43]. The dynamics of these chains has been described by reptation [56]. It seems logical that efforts be directed towards modeling the penetration by anomalous transport models and that the dissolution itself be understood through reptation.

Based on these ideas, Papanu et al. [57] suggested that when disentanglement is the controlling step, the dissolution rate be taken proportional to the radius of gyration, r_g, divided by the reptation time, t_{rep}, so that the dissolution rate R_d was expressed as

$$R_d \sim \frac{r_g}{t_{rep}} \tag{40}$$

The reptation time constant is expected to be proportional to a length squared divided by the self-diffusion coefficient, D_{self}:

$$t_{rep} \sim \frac{r_g^2}{D_{self}} \tag{41}$$

Thus,

$$R_d \sim \frac{D_{self}}{r_g} \tag{42}$$

De Gennes [58] showed that

$$r_g \sim \frac{M^{0.5}}{(1 - \upsilon_1)^{0.125}} \tag{43}$$

and

$$D_{self} \sim \frac{1}{M^{2.0}(1 - \upsilon_1)^{1.75}} \tag{44}$$

Thus, the dissolution rate was expressed as

$$R_d = \frac{K}{M^{2.5}(1 - \upsilon_1)^{1.625}} \tag{45}$$

where K is an empirical constant and M is the polymer molecular weight.

To model the swelling process, an anomalous transport model was written. The transport equation for the solvent was given by

$$\frac{\partial \upsilon_1}{\partial t} = \upsilon_2 D \frac{\partial^2 \upsilon_1}{\partial x^2} + \left(\frac{\partial \upsilon_1}{\partial x}\right)^2 \left[\upsilon_2 \frac{\partial D}{\partial \upsilon_1} - D\right] \tag{46}$$

where D is the concentration-dependent diffusion coefficient. A free volume expression was used to model the concentration dependence.

At the center of the polymer slab, a symmetry condition was used, and at the rubbery-solvent interface, the solvent concentration was equated to an equilibrium value, $\upsilon_{1,eq}$, that was estimated using thermodynamics of the swollen network. The kinetics of the glassy-rubbery interface was modeled to be governed by the stress level as

$$\frac{dR}{dt} = -K(\sigma - \sigma_c) \tag{47}$$

where σ is the total stress, σ_c is the critical stress for crazing, and K is a front factor. The critical stress, σ_c, is given by

$$\sigma_c = \gamma(T_g - T) \tag{48}$$

where γ is a constant and T_g is the glass transition temperature of the polymer.

The movement of the rubbery-solvent interface, S, was governed by the difference between the solvent penetration flux and the dissolution rate, derived earlier. An implicit Crank-Nicholson technique with a fixed grid was used to solve the model equations. A typical concentration profile of the polymer is shown in Fig. 24. Typical Case II behavior was observed. The respective positions of the interfaces R and S are shown in Fig. 25. Typical disentanglement-controlled dissolution was observed. Limited comparisons of the model predictions were made with experimental data for a PMMA-MIBK system.

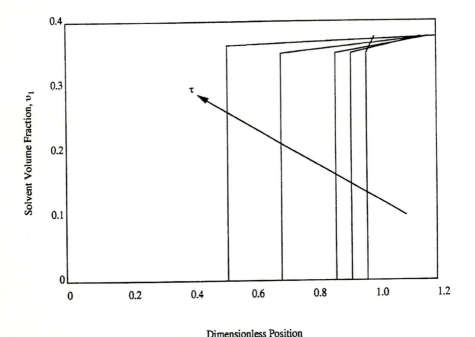

Fig. 24. Solvent concentration profiles, expressed as the solvent volume fraction, ul, and boundary locations for Case II swelling at several dimensionless times. Theoretical predictions have been adapted from the work of Papanu et al. [57], using the following parameters: $L_0 = 10^{-4}$ cm; $\chi = 0.876$; $K = 9.0 \times 10^{-9}$ cm/(s atm); $\sigma_c = 532$ atm; $V_1 = 76.5$ cm^3/mol; $D_0 = 10^{-11}$ cm^2/s; M = 200 000; T = 323 K and Pe = 1.0.

Fig. 25a, b. Positions of the glassy-rubbery (R) and the rubbery-solvent (S) interfaces as a function of dimensionless time: **a** corresponding to disentanglement limited dissolution – M = 200 000 and $\upsilon_{1,eq} = 0.64$; **b** corresponding to penetration limited dissolution – M = 50 000 and $\upsilon_{1,eq} = 0.665$. The position of the rubbery-solvent interface is denoted by the *filled circles*. Theoretical predictions have been adapted from the work of Papanu et al. [57], using the following parameters: $L_0 = 10^{-4}$ cm; $K = 9.0 \times 10^{-9}$ cm/(s atm); $\sigma_c = 532$ atm; $V_1 = 76.5$ cm^3/mol; and T = 323 K

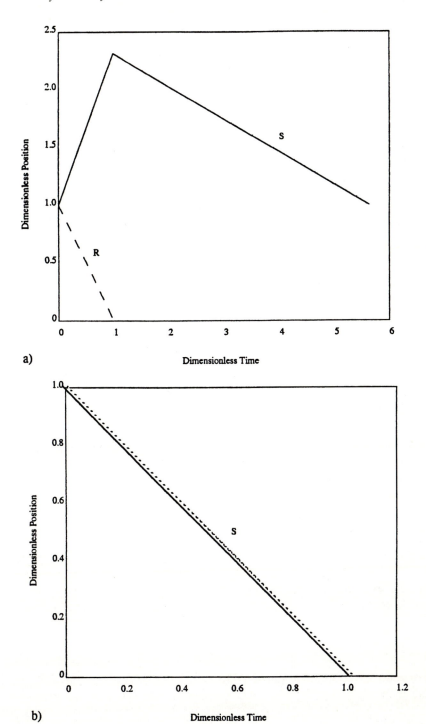

a)

b)

5.2 The Dissolution Clock Approach

Peppas et al. [59] proposed a model for the dissolution based on chain disentanglement. The radius of gyration was expressed as

$$r_g \sim \left(\frac{M}{g}\right)^{0.5} \xi \tag{49}$$

where M is the average molecular weight of the polymer, ξ is the distance between entanglements, and g is the polymer volume fraction in an entanglement subunit.

For good solvents, ξ is related to the polymer concentration as

$$\xi \sim \upsilon_2^{-0.75} \tag{50}$$

By expressing the volume of an entanglement subunit as ξ^3, the volume fraction of polymer in the entanglement subunit is

$$g = \upsilon_2 \, \xi^3 \tag{51}$$

Consequently,

$$r_g \sim M^{0.5} \, \upsilon_2^{-0.125} \tag{52}$$

The polymer self-diffusion coefficient can be expressed as

$$D_{self} \sim \frac{kT}{\eta_1 \, \xi} \left(\frac{g}{M}\right)^2 \tag{53}$$

where η_1 is the solvent viscosity. Substituting the expressions for ξ and g gave

$$D_{self} \sim \frac{kT}{\eta_1 \, \xi} M^{-2.0} \, \upsilon_2^{-1.75} \tag{54}$$

De Gennes [58] showed that

$$t_d \sim \frac{r_g^2}{D_{self}} \tag{55}$$

where t_d is the disentanglement time. Thus, the disentanglement time can be expressed as

$$t_d = K \frac{\eta_1}{kT} M^{3.0} \, \upsilon_2^{1.5} \tag{56}$$

For good solvents, ξ is proportional to $\upsilon_2^{-0.75}$. For those solvents whose χ is greater than 0.5,

$$\xi \sim \upsilon_2^{-\alpha} \tag{57}$$

Consequently,

$$t_d = K \frac{\eta_1}{kT} M^{3.0} \upsilon_2^{6\alpha-3} \tag{58}$$

where K is a function of the thermodynamic compatibility between the polymer and the solvent. Thus, a general expression for the disentanglement time was written as

$$t_d = k_d M^\alpha \upsilon_2^\beta \tag{59}$$

A model for anomalous transport [60] was extended to dissolution and the appropriate boundary condition for the rubbery-solvent interface, S, took into account the disentanglement time derived above. As in the approach of Papanu et al. [57], thermodynamics of swollen networks was used to estimate the concentration at the rubbery-solvent interface, S.

The concept of the dissolution clock was introduced. Initially, the clock is set to zero at each point in the polymer. The clock measures time only after the solvent concentration at that point reaches a critical value. When the clock time becomes equal to the disentanglement time, the polymer at that point dissolves. This is pictorially shown in Fig. 26.

The model equations were numerically solved using a finite difference technique. The predicted movement of the interfaces, R and S, with respect to time is shown in Fig. 27. The corresponding concentration profile is shown in Fig. 28. These results show the existence of three distinct stages in the dissolution process. In the first stage, there is solvent-penetration inducing swelling

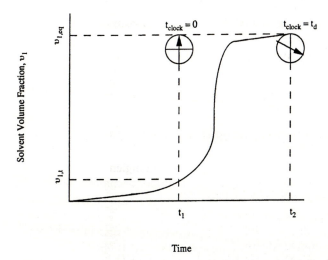

Fig. 26. Solvent volume fraction history at a spatial point. When the solvent volume fraction is equal to the critical gel volume fraction, the dissolution clock starts, t_1. When the clock time, t_2, is equal to the disentanglement time, t_d, the polymer at this point is dissolved

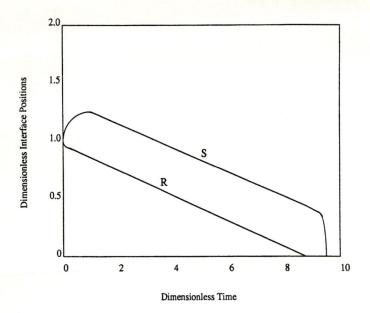

Fig. 27. Numerical predictions of the interface positions normalized with respect to the half-thickness, L_0, as functions of dimensionless time during dissolution of polystyrene in methyl ethyl ketone. The *upper curve* represents the rubbery-solvent interface while the *lower curve* represents the glassy-rubbery interface. Theoretical predictions have been adapted from the work of Peppas et al. [59], using the following parameters: $L_0 = 0.01$ cm; $M = 400\,000$; $M_c = 38\,000$; $\chi = 0.49$; $D_0 = 1.1 \times 10^{-10}$ cm^2/s; $a_d = 20$; $k_d = 2.15 \times 10^{-4}$ s; $\alpha = 2$; and $\beta = 12$

without dissolution. After an induction time, dissolution starts, and reaches a stationary phase. Then, disentanglement occurs rapidly until the polymer is completely dissolved. Experiments were performed studying PS dissolution in MEK and good agreements were obtained with the model predictions (Fig. 29).

6 Molecular Theories in a Continuum Framework

Narasimhan and Peppas [61] analyzed the dissolution mechanism of rubbery polymers by dividing the penetrant concentration field into three regimes which delineate three distinctly different transport processes. The swollen rubber was defined as the "concentrated" regime. The existence of a diffusion boundary layer adjacent to the rubbery-solvent interface, S, was proposed. This was defined as the "semi-dilute" regime. When the polymer is fully dissolved, the disentangled chains move freely in the solvent and exhibit Brownian motion ("dilute" regime, Fig. 30).

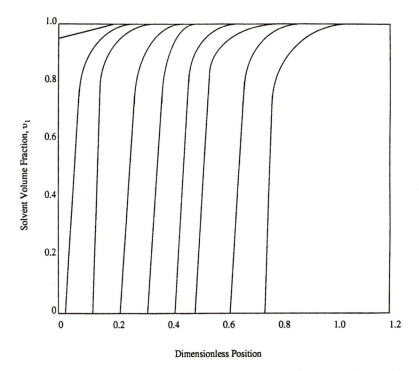

Dimensionless Position

Fig. 28. Numerical predictions of the normalized MEK volume fraction as a function of normalized position for polystyrene dissolution in MEK. Theoretical predictions have been obtained from the work of Peppas et al. [59]. The parameters are the same as in Fig. 27. The dimensionless time increases by $\Delta\theta = 0.94$ starting from the right to the left

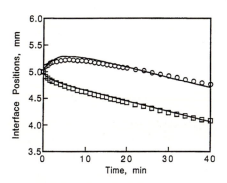

Time, min

Fig. 29. The positions of the glassy-rubbery interface (*open squares*) and the rubbery-solvent interface (*open circles*) as a function of time during monodisperse polystyrene dissolution ($\overline{M}_n = 620\,000$) in MEK at 300 K. The curves represent the predictions of the mathematical model. Reprinted with permission

The solvent penetration into the rubbery polymer was assumed to be Fickian:

$$\frac{\partial \upsilon_1}{\partial t} = \frac{\partial}{\partial x}\left(D_{12}\,\frac{\partial \upsilon_1}{\partial x}\right) \tag{60}$$

where D_{12} is the mutual diffusion coefficient between the solvent and the polymer. Appropriate initial and boundary conditions were written for Eq. (60).

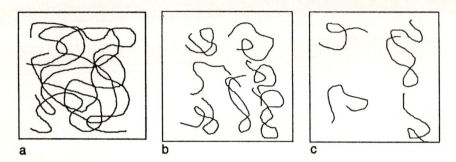

Fig. 30a–c. Disentanglement of polymer chains: **a** before dissolution starts, there is no disentanglement; this is a swellable system; **b** this depicts the onset of dissolution in the diffusion boundary layer; **c** the dissolution is complete and the disentangled chains exhibit Brownian motion in the solvent

A symmetry condition was written at the center of the polymer slab and the concentration at the interface S was estimated using thermodynamics of swollen networks [57, 62].

As the polymer chains disentangle, they move out of the gel-like phase to a liquid solution through a diffusion boundary layer of thickness δ (assumed to be constant in the present approach). The chain transport through this boundary layer is described as

$$\frac{\partial \upsilon_2}{\partial t} = \frac{\partial}{\partial x}\left(D_p \frac{\partial \upsilon_2}{\partial x}\right) - \frac{dS}{dt}\frac{\partial \upsilon_2}{\partial x} \tag{61}$$

Equation (61) is valid in the region between $x = S$ and $x = S + \delta$. In the above equation, D_p is the polymer diffusion coefficient in the solvent. The initial and boundary conditions for Eq. (61) are

$$t = 0 \quad \upsilon_2 = 0 \tag{62}$$

At the end of the boundary layer, the conventional boundary condition is

$$x = S(t) + \delta \quad \upsilon_2 = 0 \tag{63}$$

The boundary condition on the solvent side of the gel-solvent interface was written by considering that a polymer chain requires a minimum time to disentangle and move out of the gel. This minimum time is the reptation time [58]. Hence, the disentanglement rate is zero till a time equal to the reptation time elapses:

$$x = S^+(t), \quad 0 < t < t_{rep} - D_p \frac{\partial \upsilon_2}{\partial x} = 0 \tag{64}$$

After a time equal to the reptation time has elapsed, the transport of the chains at the gel-solvent interface may be disentanglement-limited or diffusion-limited. At times just greater than the reptation time, the rate of diffusion is sufficiently high and hence the flux is disentanglement-limited. Hence, the boundary condition was written as

$$x = S^+(t), \quad t > t_{rep} - D_p \frac{\partial \upsilon_2}{\partial x} = k_d \tag{65}$$

where k_d is the disentanglement rate.

As the disentanglement continues, the polymer concentration in the boundary layer increases till it reaches an equilibrium value, $\upsilon_{2,eq}$. At this instant, the diffusion rate becomes insufficient to transport the chains and hence the polymer concentration is always maintained at $\upsilon_{2,eq}$. It was proposed that an equilibrium exists between the polymer-rich gel and the polymer-lean solvent in the diffusion boundary layer. Hence the boundary condition becomes

$$x = S^+(t), \, t > t_{rep} \quad \upsilon_2^+ = \upsilon_{2,eq} \tag{66}$$

The rubbery-solvent interface S moves due to the swelling of the polymer due to solvent ingress and by subsequent chain disentanglement. This was expressed as

$$\frac{dS}{dt} = \left(D_{12} \frac{\partial \upsilon_2}{\partial x}\right)^- - \frac{D_p}{\upsilon_1^-}\left(\frac{\partial \upsilon_2}{\partial x}\right)^+ \quad S(0) = l \tag{67}$$

This completed the formulation of the moving boundary problem.

It was necessary to appropriately define the diffusion coefficients that appear in all of the preceding transport equations. It was demonstrated [61] that in dissolving systems there exists a critical solvent concentration, υ_1^*, at which the mode of mobility of the polymer chains undergoes a change. This was mathematically expressed as a change in the diffusivity and was represented as

$$\begin{aligned} D_{12} &= D_1 \quad \upsilon_1 < \upsilon_1^* \\ D_{12} &= D_2 \quad \upsilon_1 > \upsilon_1^* \end{aligned} \tag{68}$$

where

$$D_1 = D_0 \exp(a_d \upsilon_1) \tag{69}$$

where D_0 is the diffusivity of the solvent in a glassy polymer and D_2 is a "reptation" diffusion coefficient. An expression for the concentration and molecular weight dependence of D_2 was derived [61] as

$$D_2 = \frac{A}{(1 - \upsilon_1)^{1.9}} \tag{70}$$

where A is a constant that depends on polymer molecular weight, solvent viscosity and temperature, and is given as

$$A = \frac{4.8157 \times 10^{-14} \, T}{\eta_1} N^{-2.4} \tag{71}$$

To determine the critical solvent concentration, v_1^*, at which the diffusivity changes to a reptative mode, the following argument was applied. The characteristic crossover concentration for a change from a concentrated to a semi-dilute solution for polymer melts [63] is used as an initial estimate for v_1^*:

$$v_1^* = \frac{3\,N}{4\pi r_g^3} \tag{72}$$

Here, N is the number of monomers and r_g is the radius of gyration of the polymer.

The diffusion coefficient, D_p, that appears in the transport equations in the diffusion boundary layer, was defined by treating the disentangling chains in the boundary layer as Brownian spheres. Thus, a Stokes-Einstein type diffusivity arises:

$$D_p = \frac{kT}{6\pi\eta_1 r_g} \tag{73}$$

Here, η_1 is the solvent viscosity and r_g is the polymer radius of gyration. Using an exact expression for r_g [61], D_p was rewritten as

$$D_p = 1.1648 \times 10^{-14}\,\frac{T}{\eta_1\,N^{0.5}} \tag{74}$$

where T is the temperature and N is the number of monomers in the chain.

It was postulated that for a dissolving polymer, the disentanglement rate, k_d, can be given as the ratio of the radius of gyration, r_g, to the reptation time, t_{rep}:

$$k_d = \frac{r_g}{t_{rep}} \tag{75}$$

The above definition for the disentanglement rate is justified in that when chains are entangled, it is well known [58] that they require a time equal to their reptation time to disentangle. The average distance traveled by the chain during this time is equal to its primitive path [64], which in this case was approximated by using the radius of gyration of the chain. This was done as the radius of gyration is a parameter that can be experimentally measured.

Exact expressions for r_g and t_{rep} were derived [61]. The final expression for the disentanglement rate was obtained as

$$k_d = \frac{B}{(1 - v_1)^{1.9}} \tag{76}$$

where B is a constant that depends on polymer molecular weight, solvent viscosity and temperature as

$$B = \frac{4.5958 \times 10^{-9}\,T}{\eta_1}\,N^{-2.9} \tag{77}$$

This completed the formulation of the problem with all parameters defined a priori except for the "equilibrium" concentration $v_{2,eq}$ in the diffusion boundary layer. Owing to lack of any prior knowledge of this concentration, this was treated as a parameter.

The model equations (Eqs. (60) and (61)) were solved numerically using an implicit finite difference technique. Typical profiles for the solvent volume fraction as a function of position and time in the rubber and the polymer volume fraction in the diffusion boundary layer are shown in Figs. 31 and 32 respectively. Other features of the simulation are the prediction of the temporal evolution of the rubbery-solvent interface and the mass fraction of the polymer dissolved as a function of time (Figs. 33 and 34). The simulations showed that the dissolution could be either disentanglement or diffusion controlled depending on the polymer molecular weight and the thickness of the diffusion boundary layer.

This approach was furthered by Narasimhan and Peppas [65] to understand the dissolution of a glassy polymer in a solvent. The concentration field was divided into three regimes similar to the previous approach [61]. A linear irreversible thermodynamics argument was used to obtain the volumetric flux of the solvent as

$$v_1 V_1 = -\left[D_{12} \frac{\partial v_1}{\partial x} + \frac{D_{12} \bar{V}_1 v_1}{RT(1 - v_1)(1 - 2\chi v_1)} \frac{\partial \pi}{\partial x} \right]. \tag{78}$$

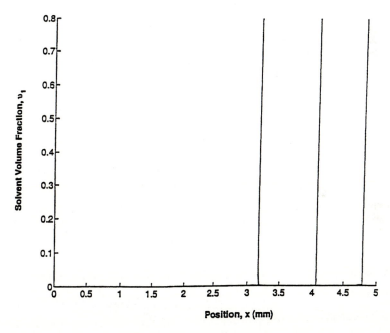

Fig. 31. Solvent volume fraction, v_1, as a function of position, x. The polymer molecular weight was $\bar{M}_n = 40\,000$ and $t_{rep} = 1000$ s. The position x = 0 is the center of the slab. The time increment starting from the first curve on the right is $\Delta t = 1000$ s

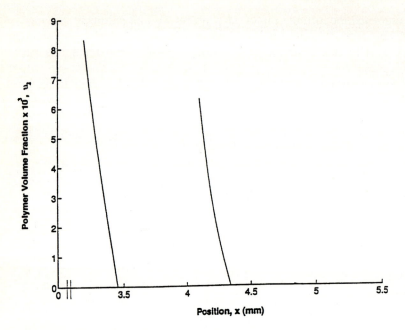

Fig. 32. Polymer volume fraction, υ_2, in the diffusion boundary layer, as a function of position, x. The polymer molecular weight was $\bar{M}_n = 40\,000$ and $t_{rep} = 1000$ s. The position x = 5 mm represents the initial slab thickness. The time increment starting from the first curve on the right is $\Delta t = 1000$ s

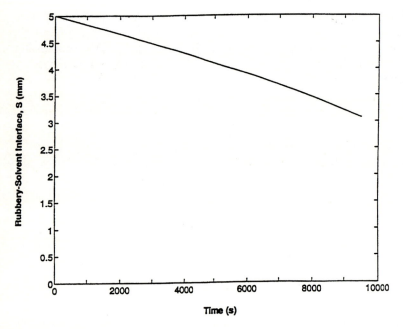

Fig. 33. The position of the rubbery-solvent interface, S, as a function of dissolution time. The polymer molecular weight was $\bar{M}_n = 40\,000$ and $t_{rep} = 1000$ s

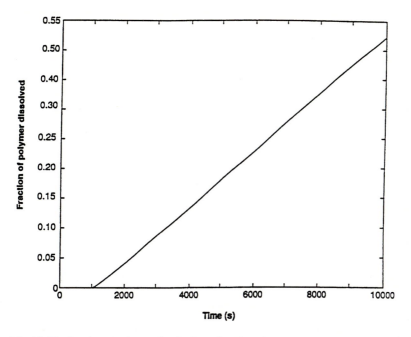

Fig. 34. The fraction of polymer dissolved as a function of time. The polymer molecular weight was $M_n = 40\,000$ and $t_{rep} = 1000$ s

Thus, the solvent flux was expressed as a sum of contributions from diffusive and osmotic pressure terms. The osmotic pressure in the solvent flux expression depends upon the viscoelastic properties of the polymer. The relationship between the osmotic pressure and the stresses within the polymer was derived by writing momentum balances as

$$\frac{\partial \pi}{\partial x} = \frac{\partial \sigma_{xx}}{\partial x} \tag{79}$$

where σ_{xx} is the normal stress. A Maxwell model was used as the constitutive equation for the polymer and the set of governing equations for the concentrated regime were thus obtained as

$$\frac{\partial v_1}{\partial t} = \frac{\partial}{\partial x}\left[D_{12}\frac{\partial v_1}{\partial x}\right] + \frac{\partial}{\partial x}\left[\frac{D_{12}\,\bar{V}_1\,v_1}{RT(1-v_1)(1-2\chi v_1)}\frac{\partial \sigma}{\partial x}\right] \tag{80}$$

$$\frac{\partial \sigma_{xx}}{\partial t} = -\frac{\sigma_{xx}}{(\eta/E)} + \frac{E}{(1-v_1)^2}\frac{\partial v_1}{\partial t} \tag{81}$$

where χ is the solvent-polymer interaction parameter. The transport in the semi-dilute regime was modeled as before [61] and the resulting set of equations was numerically solved using an implicit finite difference technique. The features of the simulation are shown in Figs. 35–38. Figure 35 shows the solvent volume

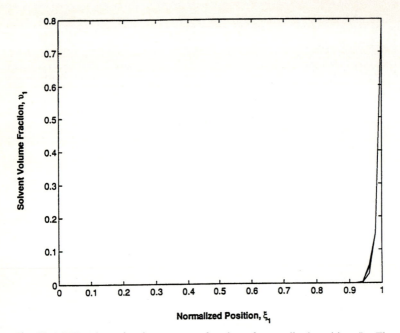

Fig. 35. MEK volume fraction, υ_1, as a function of normalized position, ξ_1. The polystyrene molecular weight was $\bar{M}_n = 52\,000$. The position $\xi_1 = 0$ is the center of the slab. The time increment starting from the first curve on the right is $\Delta t = 1000$ s

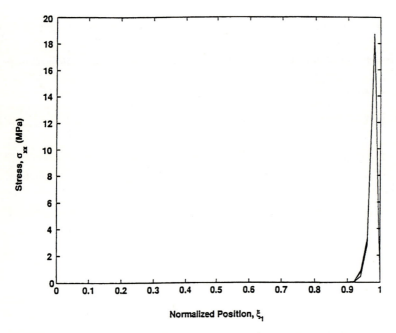

Fig. 36. Internal stress, σ_{xx}, as a function of normalized position ξ_1. The polystyrene molecular weight was $\bar{M}_n = 52\,000$. The position $\xi_1 = 0$ represents the center of the slab. The time increment starting from the first curve on the right is $\Delta t = 1000$ s

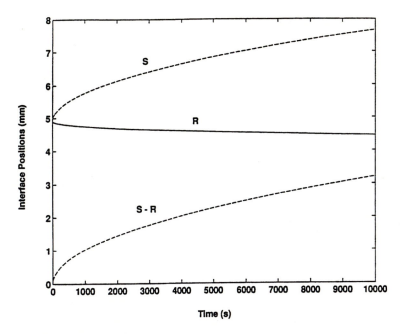

Fig. 37. The rubbery-solvent interface (S), the glassy-rubbery interface (R), and the gel layer thickness (S-R) as a function of dissolution time. The polystyrene molecular weight was $\bar{M}_n = 52\,000$

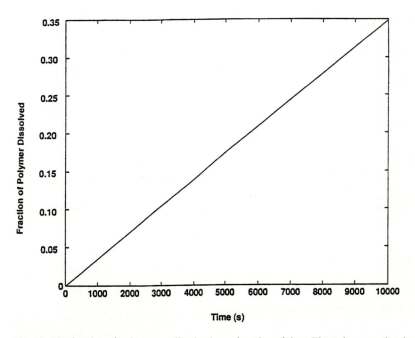

Fig. 38. The fraction of polystyrene dissolved as a function of time. The polymer molecular weight was $\bar{M}_n = 52\,000$

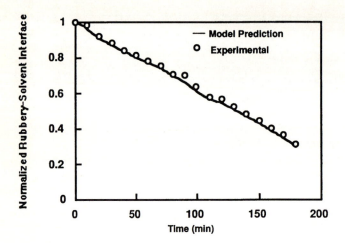

Fig. 39. The position of the rubbery-solvent interface (*open circles*) as a function of time during monodisperse poly(ethylene glycol) dissolution ($\bar{M}_n = 35\,000$) in water at 300 K. The curve represents the predictions of the mathematical model

fraction in the polymer as a function of time and position for polystyrene ($\bar{M}_n = 52\,000$) dissolved in methyl ethyl ketone. Figure 36 shows the internal stress profile in the polymer. It is instructive to note that the stress value in the rubber is of the order of 20 MPa and this could be used to characterize the mechanism of dissolution. The positions of the interfaces R and S as a function of time are shown in Fig. 37. Integrating the concentration profiles, the mass fraction of polymer dissolved as a function of time was obtained (Fig. 38). The effect of the polymer molecular weight and the diffusion boundary layer thickness on the dissolution mechanism was investigated for polystyrene dissolution in methyl ethyl ketone. The results showed that upon increasing the polymer molecular weight, the dissolution became disentanglement-controlled. Decrease in the diffusion boundary layer thickness led to a shift in the dissolution mechanism from disentanglement-control to diffusion-control.

Experiments were performed to study the dissolution of poly(ethylene glycol) (PEG) in water and the position of the rubbery-solvent interface as a function of time was observed using a photographic technique. Simulations were performed for dissolution of PEG in water and the comparison with the experimental data is good (Fig. 39).

7 Discussion and Analysis

The various approaches to model polymer dissolution have accounted for stress relaxation, viscoelasticity of the polymer, disentanglement of polymer chains, anomalous transport of solvent, chain reptation, and external mass transfer limitations. In the process, some questions have been left unanswered.

The model of Tu and Ouano [43] identified a polymer "disassociation" rate as a key parameter. However, they failed to predict quantitatively this rate from the molecular properties of the polymer and the solvent as they treated the disassociation rate as a model parameter.

The phenomenological disengagement dynamics models [46–48] assumed that the solvent concentration at the solvent-polymer interface is independent of the solvent concentration history. There are parameters used that cannot be obtained from experiments. So the physical origin of such parameters is not very clear. In addition, there is a lot of empiricism in the approach.

Although the external mass transfer controlled polymer dissolution approach [51] may be intuitive, experiments have indicated that external mass transfer effects are insignificant. Papanu et al. [40] showed that for dissolution of PMMA in MIBK, vigorous agitation of the solvent increased the dissolution rate by only 15% relative to that for a stagnant solvent. Also, since the chain disentanglement mechanism was not considered, these models fail to explain the swelling time needed before dissolution.

The stress relaxation approach [53, 55] was able to account for both solvent penetration and actual polymer dissolution. But it involved too many assumptions and several molecular parameters had to be determined by experiment. In addition, no effort was made to compare the modeling results to experimental observations.

The reptation idea could account for the effect of polymer molecular weight and solvent concentration on the dissolution rate. However the key concentration identified in this approach [57, 59] is one at the surface. This implies independence of the solvent concentration history. This may not be true as the disentanglement of polymer chains does not commence till the local solvent concentration is greater than a critical value at which the local glass transition temperature is lowered below the experimental temperature so that the glassy polymer changes into a gel. Also, all of the efforts discussed so far failed to take into account the effect of the viscoelastic properties of the polymer on the dissolution mechanism.

The approach of using molecular theories in a continuum framework [61, 65] seems to yield useful insight into the understanding of the physics of the problem. In addition, the effect of the viscoelastic nature of the polymer has been taken into account. Comparisons with experimental data appear promising. The effect of polydispersity on the dissolution mechanism is another aspect that has not been addressed by any of the above contributions.

In conclusion, all the above approaches have addressed various issues that are important in the polymer dissolution phenomenon. The comparisons with experiments are still at a qualitative level. Experiments that will establish quantitative validation of the models are necessary and this is where the future thrust should be in this area.

Acknowledgments. This work was supported by Grant No. CTS-92-12482 from the National Science Foundation.

8 References

1. Cooper WJ, Krasicky PD, Rodriguez F (1986) J Appl Polym Sci 31: 65
2. Iwayanagi T, Ueno T, Nonogaki S, Ito H, Willson CG (1988). In: Electronic and photonic applications of polymers, ACS Symposium Series, Vol 218, ACS, Washington, DC, pp 109–224
3. O'Brien MJ, Soane DS (1989). In: Microelectronics processing: chemical engineering aspects, ACS Symposium Series, Vol 221, ACS, Washington, DC, pp 325–376
4. Parsonage EE, Peppas NA (1987) Brit Polym J 19: 469
5. Parsonage EE, Peppas NA, Lee PI (1987) J Vac Sci Technol B5: 538
6. Colombo P, Gazzaniga A, Sangalli ME, La Manna A (1987) Acta Pharm Technol 33: 15
7. Colombo P, Catellani PL, Peppas NA, Maggi L, Conte U (1992) Int J Pharm 88: 99
8. Conte U, Colombo P, Gazzaniga A, Sangalli ME, La Manna A (1988) Biomaterials 9: 489
9. Harland RS, Gazzaniga A, Sangani ME, Colombo P, Peppas NA (1988) Pharmaceut Res 5: 488
10. Lee PI (1987) In Controlled Release Technology: Pharmaceutical Applications, ACS Symposium Series, Vol 348, ACS, Washington, DC, pp 71–83
11. Pham AT, Lee PI (1993) Proceed Intern Symp Control Rel Bioact Mater 20: 220
12. Tsay CS, McHugh AJ (1990) J Polym Sci Polym Phys Ed 28: 1327
13. Yilmaz L, McHugh AJ (1988) J Appl Polym Sci 35: 1967
14. Kesting RE (1993) Polymeric Gas Separation Membranes, Wiley, New York, NY
15. Nauman EB, Lynch JC (1993) US Patent 5,198,471
16. Nauman EB, Lynch JC (1994) US Patent 5,278,282
17. Dammel RR (1993) Polym Mater Sci Eng Proceed 68: 49
18. Yeh TF, Reiser A, Dammel RR, Pawlowski G, Roeschert H (1993) Macromolecules 26: 3862
19. Huang JP, Kwei TK, Reiser A (1989) Macromolecules 22: 4106
20. Huang JP, Pearce EM, Reiser A, Kwei TK (1989) Polym Mater Sci Eng Proceed 60: 280
21. Wadke DA, Reier GE (1972) J Pharm Sci 61: 868
22. Koida Y, Takahata H, Kobayashi M, Samejime M (1987) Chem Pharm Bull 35: 1538
23. Cooney DO (1972) AIChE J 18: 446
24. Ozturk SS, Palsson BO, Dressman JB (1988) Pharmaceut Res 5: 272
25. Ozturk SS, Palsson BO, Donohue B, Dressman JB (1988) Pharmaceut Res 5: 550
26. Ueberreiter K (1968) In Diffusion in Polymers, Academic Press, New York, NY
27. Raptis G (1965) PhD Dissertation, Free University of Berlin, Berlin
28. Ueberreiter K, Asmussen F (1962) J Polym Sci 57: 187
29. Ouano AC, Carothers JA (1980) Polym Eng Sci 20: 160
30. Rodriguez F, Krasicky PD, Groele RJ (1985) Solid State Technol 28: 125
31. Cooper WJ, Krasicky PD, Rodriguez F (1985) Polymer 26: 1069
32. Krasicky PD, Groele RJ, Jubinsky JA, Rodriguez F (1987) Polym Eng Sci 27: 282
33. Krasicky PD, Groele RJ, Rodriguez F (1987) Chem Eng Comm 54: 279
34. Groele RJ, Rodriguez F (1989) J Coatings Tech 66: 55
35. Rao V, Kosbar LL, Frank CW, Pease RFW (1991) Proc SPIE 1672: 214
36. Rao V, Kosbar LL, Frank CW, Pease RFW (1994) In Polymers for Microelectronics, ACS Chemistry Series, ACS, Washington, DC, pp 220–234
37. Manjkow J, Papanu JS, Hess DW, Soane DS, Bell AT (1987) J Electrochem Soc 134: 2003
38. Manjkow J, Papanu JS, Hess DW, Soane DS, Bell AT (1987) J Appl Phys 62: 682
39. Papanu JS, Hess DW, Soane DS, Bell AT (1989) J Electrochem Soc 136: 1195
40. Papanu JS, Hess DW, Soane DS, Bell AT (1989) J Electrochem Soc 136: 3077
41. Papanu JS, Hess DW, Soane DS, Bell AT (1990) J Appl Polym Sci 39: 803
42. Drummond RK, Boydston GL, Peppas NA (1990) J Appl Polym Sci 39: 2267
43. Tu YO, Ouano AC (1977) IBM J Res Develop 21: 131
44. Tu YO (1977) Quart Appl Math 35: 269
45. Ouano AC, Tu YO, Carothers JA (1977) In Structure–Solubility Relationships in Polymers, Academic Press, New York, NY

46. Devotta I, Ambeskar VD, Mandhare AB, Mashelkar RA (1994) Chem Eng Sci 49: 645
47. Ranade VV, Mashelkar RA (1995) AIChE J 41: 666
48. Devotta I, Badiger MV, Rajamohanan PR, Ganapathy S, Mashelkar RA (1995) Chem Eng Sci 50: 2557
49. Zielinski JM, Duda JL (1992) AIChE J 38: 405
50. Astarita G, Sarti GC (1978) Polym Eng Sci 18: 388
51. Lee PI, Peppas NA (1987) J Contr Rel 6: 207
52. Lee HR, Lee YD (1991) Chem Eng Sci 46: 1771
53. Brochard F, de Gennes PG (1983) Phys Chem Hydrodynamics 4: 313
54. Doi M, Edwards SF (1986) The Theory of Polymer Dynamics, Oxford University Press, Oxford
55. Herman MF, Edwards SF (1990) Macromolecules 23: 3662
56. de Gennes PG (1971) J Chem Phys 55: 571
57. Papanu JS, Soane DS, Bell AT, Hess DW (1989) J Appl Polym Sci 38: 859
58. de Gennes (1979) Scaling Concepts in Polymer Physics, Cornell University Press, Ithaca, New York
59. Peppas NA, Wu JC, von Meerwall, ED (1994) Macromolecules 27: 5626
60. Wu JC, Peppas NA (1993) J Polym Sci Polym Phys Ed 31: 1503
61. Narasimhan B, Peppas NA (1996) J Polym Sci Polym Phys Ed, 34: 947
62. Flory PJ (1953) Principles of Polymer Chemistry, Cornell University Press, Ithaca, New York
63. des Cloizeaux J, Jannink G (1989) Polymers in Solution: Their Modeling and Structure, Oxford University Press, Oxford
64. Edwards SF (1982) Proc Royal Soc London A385: 267
65. Narasimhan B, Peppas NA (1996) Macromolecules, 29: 3283

Editor: S. Edwards
Received: March 1996

Author Index Volumes 101-128

Author Index Vols. 1-100 see Vol. 100

Subject Index

Springer
and the
environment

At Springer we firmly believe that an
international science publisher has a
special obligation to the environment,
and our corporate policies consistently
reflect this conviction.
We also expect our business partners –
paper mills, printers, packaging
manufacturers, etc. – to commit
themselves to using materials and
production processes that do not harm
the environment. The paper in this
book is made from low- or no-chlorine
pulp and is acid free, in conformance
with international standards for paper
permanency.

Printing: Saladruck, Berlin
Binding: Buchbinderei Lüderitz & Bauer, Berlin

RETURN TO ➡

CHEMISTRY LIBRARY
100 Hildebrand Hall • 642-3753

LOAN PERIOD 1	2	3
4	5	2 HOUR

~~ALL BOOKS MAY BE RECALLED AFTER 7 DAYS~~
~~Renewable by telephone~~

DUE AS STAMPED BELOW

~~AUG 10 2001~~		
AUG 15 2003		